ORIGINS

The Day the Earth was Smashed like a Cup

I0476193

ISBN 13: 978-1511761123

ISBN 10: 1511761121

http://arkpublishing.co.uk/

Bernard Paul Badham

*'A Thousand years is like a day unto the Lord,
and a day a thousand years.'*

EINSTEIN

'Time and Space are Relative.'

CONTENTS

Foreword

When we look back in time at our own recorded history, archaeological or written in stone or in religious and mythological texts, we have a simplified and sterile view of our origins. Truth may be stranger than fact. Our ancestors interpreted terrestrial and astronomical events through the eyes of their own misunderstandings and religious and cultural beliefs, today, with science and logic, these events may be interpreted more correctly, but it is important to note that myth always has some basis in fact.

The origins of life, human kind, civilization and the validity of mythical historical events need to be re-examined through the looking-glass of science, archaeology, astrophysics, mathematics and of course sacred ancient written accounts and legends.

When we look up at the heavens, the planets, stars and galaxies, we interpret their origins based on modern scientific evidence, but this evidence may be interpreted to give a different picture to what we traditionally understand concerning our origins, for instance in the Theory of the Big Bang. One of the most puzzling things to be examined by this book is the origin of water on our planet Earth, the vast oceans of water, essential to life should not be there. New evidence about the geology of our Earth is coming to light which may provide a surprising conclusion about its origins and eventual consequences in how it affected human development and the origins of civilization.

This book is made up of short essays on important topics which affect our origins, it is best read in the order presented, but may be used as a 'dip in read' according to what interests you.

Introduction

The purpose of this book is to examine the scientific evidence and religious, cultural and historical accounts concerning our origins, the origins of the universe, space, time and matter, life itself and the birth of civilization. Today we have a distinct separation between what we believe (faith) and science, for our ancestors this distinction did not exist, that which they saw and experienced was interpreted by what they believed, but in the beginning this knowledge experience led to belief. The ancient Egyptians saw a harmony between their everyday experiences and what they believed, one was entwined with the other. The world around them needed explanation and so they interpreted their experience of daily phenomena in a way they could understand. The great giver of life, warmth and light, which traversed the heavens each day was interpreted as a living, powerful being, the sun god Ra, it was therefore no surprise that they worshipped the sun and prayed each morning before sunrise for its return to deliver them from the coldness and darkness of night.

Their interpretation of the sun's disappearance under the earth-sky horizon in the west each evening and its rise each morning in the eastern horizon gave birth to the beliefs of the underworld and the resurrection of life. They interpreted what they saw and experienced in the only way they could, by applying their own human experiences of life, birth and death to the 'inanimate' universe around them. Today science has a clear distinction between the living (animate) and the non-living (in-animate), the ancients did not have such distinctions, the whole universe was animate, and alive.

As modern, educated humans we have a simplistic view of our ancestors, we see their culture and beliefs as primeval, pagan even and misinformed, but in a way they had the advantage of being 'in touch with nature' and their thinking was not limited by educational dogma, they were free thinkers. One of the greatest scientists of our modern times, Albert Einstein, said that he had to undo, put aside that which he had been taught in order to free himself to see and explain scientific phenomena with unfettered thinking.

This view of a living, breathing universe is evidenced in their mythological beliefs and in their scientific, medical and astronomical knowledge. Their hieroglyphic written language for instance refers to inanimate objects as either feminine or masculine.

ra, the sun (masculine)

akhet, the horizon (feminine)

It is not surprising that the sun, as in the male sun god Ra was referred to as masculine. The horizon as feminine may have come from the fact that the sky goddess Nut swallowed the sun each evening and gave birth to it each morning.

Modern science is confined to time frames, whether historical, geological or astronomical, time plays a fundamental role in the perception of our origins and the origin and evolution of the universe and life itself. But the concept of time is as elusive as the definition and true understanding of the nature of energy, matter and light. Many arguments rage between traditional scientists, atheists and the like 'the evolutionists' and religious thinkers, the 'creationists' concerning our origins, much of this debate falters around historical time frames, but maybe we should simply search for the truth about such matters, therefore, we need to re-examine the evidence, unfettered and unbiased and without emotion, if that were possible. We need to think freely, but in the light of modern historical, geological and astronomical scientific evidence. This dualistic approach may reveal some interesting conclusions. Let us start with the concept of time.

Time is a constant in a measured frame of reference, clocks 'tick-tock' constantly, but time can vary between one frame of reference and another in our universe and even over time itself.

Time is one of the hardest things to explain, we understand it until we come to explain it, but time and particularly astronomical, geological and historical time scales are absolutely key in understanding our origins:

1. Modern Scientific Timeline from the Big Bang to the Birth of Civilization

Scientifically our universe has been dated to being 13.8 billion years old, this timing comes from the Big Bang and Expansion of the Universe theory which we will look at in detail. The age of our Solar System and the formation of the Earth are estimated to be 4.5 billion years. The emergence of life on our planet in the form of simple (prokaryotic) cells 3.6 billion years, complex (eukaryotic) cells 2 billion years, multicellular life 1 billion years, simple animals 600 million years, land plants 475 million years, mammals 200 million years, primates 60 million years, human predecessors (genus *Homo*) 2.5 million years and modern humans (*Homo Sapiens*), 200,000 years:

Homo Sapiens (Africa) 200,000 years ago

Human colonization of continents 50,000 a

Giant lizards die out 40,000 a

Woolly rhinoceros becomes extinct 15,000 a

Bears, giant sloth and all equidae (in North America) die out 11,000 a

Late Glacial Maximum 13,000 - 10,000 a

After an incredible time span of nearly 200,000 years, then an only then, do we see the emergence of civilization? This unbelievably long time gap needs explanation. Traditionally we are told that we wandered the Earth for generation upon generation as nomads, hunter gatherers, with no thought of permanent settlements or farming of the land, but historical and mythological records tell us otherwise, of great kings and kingdoms, were our ancestors fantasying about their own history?

I think not, in a later chapter we will examine some of these so called myths about pre-civilizations, in the meantime let us continue to examine the traditional theories of our origins.

Start of the Modern Warm Period 8,000 BC

Agricultural Revolution 8,000 - 5000 BC
(In the Fertile Crescent)
Egypt - Phoenicia - Mesopotamia

Humans began the systematic husbandry of plants and animals. Agriculture advanced, and most humans transitioned from a nomadic to a settled lifestyle as farmers in permanent settlements.

The Fertile Crescent from where civilization emerges includes Egypt, Phoenicia and Mesopotamia, the land in and around the Tigris and Euphrates rivers. The modern-day countries with significant territory within the Fertile Crescent are Iraq, Kuwait, Syria, Lebanon, Jordan, Israel, Cyprus, and Egypt, besides the south-eastern fringe of Turkey and the western fringes of Iran.

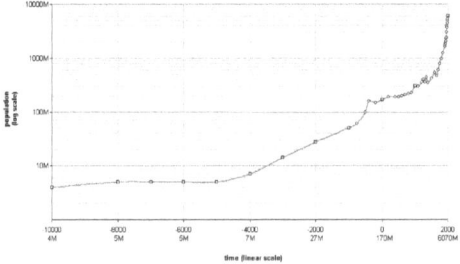

World population growth since 10,000 BC to present

Note: using highly conservative calculations for population growth from an original pair to today's numbers yields that our origins date back only 5000 years!

The region is often called the cradle of civilization; it saw the development of many of the earliest human civilizations. Some of its technological inventions are writing, glass, the wheel and the use of irrigation. The earliest known western civilizations manifestly arose and flourished using the water supplies and agricultural resources available in the Fertile Crescent. They were not necessarily the first or the only source of civilization. Moreover, plants and animals were not domesticated there but in the surrounding areas.

Ancient recorded history begins with the invention of writing, however, the roots of civilization reach back to the period before the invention of writing. Prehistory begins in the Paleolithic Era, or 'Early Stone Age,' which is followed by the Neolithic Era, or New Stone Age, and the Agricultural Revolution in the Fertile Crescent.

The Agricultural Revolution marked a change in human history, as humans began the systematic husbandry of plants and animals. Agriculture advanced, and most humans transitioned from a nomadic to a settled lifestyle as farmers in permanent settlements. Nomadism continued in some locations, especially in isolated regions.

As farming developed, grain agriculture became more sophisticated and prompted a division of labor to store food between growing seasons. Labor divisions then led to the rise of a leisured upper class and the development of cities. The growing complexity of human societies necessitated systems of writing and accounting. Many cities developed on the banks of lakes and rivers; as early as 3000 BC some of the first prominent, well-developed settlements had arisen in Mesopotamia, on the banks of Egypt's River Nile, and in the Indus River valley. Similar civilizations probably developed along major rivers in China, but archaeological evidence for extensive urban construction there is less conclusive.

Thus begins the origins of our civilization.

2. In the Beginning - The Trouble with Time

Time is essential to understanding our origins, for out of universal time comes astronomical, geological and historical time lines. As stated above our universe is said to have come into existence 13.8 billion years ago and our Earth Solar System and Sun, around 4.5 billion years ago. Our whole dating system for the age of the universe comes from one fundamental assumption that time itself is constant. This is not the case. So where did the time scale of 13.8 billion years for the Big Bang and the age and origins of the universe come from? The answer of course is from modern scientific evidence. The main source of evidence for this is from our modern view of the present universe, that it is expanding, for after the Big Bang where matter, light, space and time came into existence, the universe expanded to its present size and state.

Evidence for the Expansion of the Universe and the Big Bang

The Big Bang theory is the prevailing cosmological model for the early development of the universe. The key idea is that the universe is expanding. Consequently, the universe was denser and hotter in the past. Moreover, the Big Bang model suggests that at some moment all matter in the universe was contained in a single point, which is considered the beginning of the universe. Modern measurements place this moment at approximately 13.8 billion years ago; which is therefore considered the age of the universe. After the initial expansion, the universe cooled sufficiently to allow the formation of subatomic particles, including protons, neutrons, and electrons. Though simple atomic nuclei formed within the first three minutes after the Big Bang, thousands of years passed before the first electrically neutral atoms formed. The majority of atoms that were produced by the Big Bang were hydrogen, along with helium and traces of lithium. Giant clouds of these primordial elements later coalesced through gravity to form stars and galaxies, and the heavier elements were synthesized either within stars or during supernovae.

The two main sources of evidence for the Big Bang and the expansion of the universe are:

1. The Red Shift of light from distant galaxies:

In 1929, Edwin Hubble discovered that the distances to far away galaxies were strongly correlated with their redshifts. Hubble's observation was taken to indicate that all distant galaxies and clusters have an apparent velocity directly away from our vantage point: that is, the farther away, the higher the apparent velocity, regardless of direction.

Assuming that we are not at the center of the universe,

the only remaining interpretation is that all observable regions of the universe are receding from each other - the universe is expanding.

2. Cosmic Microwave Background Radiation:

The cosmic microwave background radiation (CMBR) is the thermal radiation assumed to be left over from the 'Big Bang' of cosmology. CMBR is also referred to as 'relic radiation.' The CMBR is a cosmic background radiation that is fundamental to observational cosmology because it is the oldest light in the universe. Sensitive radio telescopes show a faint background glow in space, almost exactly the same in all directions; this radiation is not associated with any star, galaxy, or other object. This glow is strongest in the microwave region of the radio spectrum.

The CMBR is well explained as radiation left over from an early stage in the development of the universe, and its discovery is considered a *landmark test of the Big Bang model of the universe.* When the universe was young, before the formation of stars and planets, it was denser, much hotter and filled with a uniform glow from a white-hot fog of hydrogen plasma. As the universe expanded, both the plasma and the radiation filling it grew cooler. When the universe cooled enough, protons and electrons combined to form neutral atoms. These atoms could no longer absorb the thermal radiation, and so the universe became transparent. The temperature of this background radiation is only just above absolute zero (0 kelvin, -273 Celsius), its peak microwave wavelength temperature is 2.73 kelvin.

It is assumed therefore that if today the universe is expanding, less dense and cold, that in the past it must have been contracted, more dense and hotter. This is an extrapolation of the current nature of the universe back in time and assumes that the rate of expansion was a 'constant.' One of the first things I teach my physics students when plotting acquired experimental data is to not extrapolate their graphs beyond the data, for that is not science, but assumption.

Extrapolation of the expansion of the universe backwards in time using general relativity yields an infinite density and temperature at a finite time in the past. This singularity signals the breakdown of general relativity. How closely we can extrapolate towards the singularity is debated. This singularity is sometimes called 'the Big Bang', but the term can also refer to the early hot, dense phase itself, which can be considered the 'birth' of our universe.

In our expanding universe model, it is space itself which is expanding, not galaxies themselves. Imagine sticking small beads on the surface of a rubber balloon, as the balloon is inflated the beads move further away from each other. This expansion of space leads to the increasing wavelength of light producing a shift in its spectrum towards the longer wavelength red end, this is the red shift.

Apart from the above assumptions, that time is constant and that the expansion was 'constant', there is another assumption that gravity between masses in the universe was constant, i.e. Newton's gravitational constant (G) is actually a constant over time and expansion and since gravity is the known breaking force on expansion then the way gravity behaves over time is important.

One thing we know for sure is that our universe exists and has been around for a long time, but is it really expanding? Is the universe really 13.8 billions of years old? Are there any other explanations for CMBR and Red Shift? And over the history of the universe is time and gravity truly constant?

3. The Constant Nature of Time and the Speed of Light

We all understand time; it ticks by with perfect order. Our ancient ancestors recorded the passage of time by the movements of the heavenly bodies, particularly the sun and moon. One rotation of the Earth gives us one day divided into 24 hours; the seasonal changes governed by the orbit of the Earth around the sun gave us the year. Today modern clocks can measure time extremely accurately and with a precision of trillionths of a second. But what governs the phenomenon of time and why is it regarded as constant?

Until Einstein we had a cozy view of the universe, space and of time, but in the early twentieth century he caused a major upset in the scientific community, the main reason being his Special and General Relativity theory which states that space and time are not constant, but relative. Space and time can vary depending on position, speed and gravity.

Special Relativity: at terrestrial speeds time is seemingly unaffected, but as one approaches the speed of light time slows down exponentially. Theoretically at the speed of light, time stops.

Note: also an object's mass increases exponentially as the speed of light is approached.

General Relativity: in a gravitational field time slows down and in extreme gravity fields such as at the event horizon of a Black Hole, time stops.

This all being relative to a stationary observer: in the SR case; and to an observer outside a gravity field in the GR case. Not only does position and speed affect time, it also affects space length. To a stationary observer, a spaceship approaching near light speed would appear to contract in length. Einstein's relativity theories regard these effects as space and time dilation. Thus from his theory came the concept space-time.

The most weird, and wonderful, thing to come out of relativity is that anyone on board a spaceship approaching the speed of light or near the event horizon of a black hole would not notice, or be able to measure, these space and time dilations. If measured under these extreme conditions, the speed of light would still have the same value (c) and time would appear to tick normally, it is only an outside observer who would notice that clocks inside the spaceship run slower. Also for those on board, a meter ruler would still appear the correct length. These relativistic effects Einstein referred to has giving harmony to the universe, no matter where you are or at what speed you are travelling everything appears normal, time, length and the speed of light. Space and time dilation are not just relativistic effects, but are very real and measurable, and proven by scientists across the world over the last century.

This leaves the question, why? What is actually happening to cause space and time dilation? How do these effects affect the evolution of an expanding universe and its apparent age?

4. The General Theory of Relativity

Einstein describes space and time as space-time, space and time are inseparable. It is the distortion of this space-time matrix which results in space and time dilation. In GTR a gravitational field is a distortion of the space-time matrix, the matrix being the mathematical geometry of space-time. This matrix has 4 dimensions, three of space (3D) and one of time. Any mass causes a distortion of this space-time matrix, the larger the mass the greater the distortion and hence the greater the space and time dilation.

This raises another question: how can space (and time) be distorted since the vacuum of space contains nothing to distort? If space has substance, then the distortion of the fabric substance of space makes sense, but herein lies a century raging augment concerning whether what we see as empty space is not empty at all. Space is filled with an invisible and almost undetectable substance, primarily this inferred substance of the fabric of space is called the ether, a phenomenon which almost all modern scientists refuse to believe exists, for there is 'no evidence' of its existence. Einstein referred to this particular problem in his 1920 Leyden lecture:

'...Recapitulating, we may say that according to the general theory of relativity *space is endowed with physical qualities*; in this sense, therefore, there exists an ether. According to the general theory of relativity space without ether is unthinkable; for in such space there not only would be no propagation of light, but also no possibility of existence for standards of space and time (measuring-rods and clocks), nor therefore any space-time intervals in the physical sense. But this ether may not be thought of as endowed with the quality characteristic of *ponderable media*, as consisting of parts which may be tracked through time. The idea of motion may not be applied to it.'

If the ether exists, then the GTR makes sense, space has substance and physical properties which can be distorted by a gravitating mass. With regards to the physical nature of the ether, Einstein suggests that it is not made of substance (ponderable matter) as we know it; it is invisible and 'undetectable.'

In Einstein's GTR it can no longer be accepted that the curvature of space-time is purely a mathematical concept, space must have physical substance and properties which affect the speed of light and hence time. The effects of this 'imponderable matter' must also be related to the known effects of STR: this physical substance of space must affect objects travelling at speed. The *physical nature of space* must be responsible for the observable and measurable effects in both General Relativity *and* Special Relativity?

Recapitulating:

General Theory of Relativity

1. In the vacuum of space with a zero gravitational field time ticks away 'normally,' length is not contracted and the speed of light is (c).

2. In the vacuum of space with a gravitational field time is slowed down, length is contracted and the speed of light is slowed down, less than (c)

> *These effects seen in 2 are by an observer in 1,*
> *but an observer in 2 would not see these effects.*

5. Special Theory of Relativity

In the vacuum of space for an object travelling at near light speeds, time slows down, length is contracted and its mass increases, this is because as the craft approaches the speed of light, the energy-density within the craft moving through the space vacuum approaches infinity, therefore, length and time approach zero and space-vacuum drag (mass) approaches infinity.

The effects in GTR and STR must be related and as Einstein stated for GTR, these effects have their cause anchored in the physical nature of space.

The following discuses GR, for in order to understand special relativity[*], we need to be able to explain the effects in general relativity. It is clear from both theories that somehow length, time, mass and the speed of light are interwoven with each other and with the physical nature of space, the ether. Without regard at this point to the physical composition of the substance (imponderable matter) of the ether, can these relativistic effects be linked and explained? Classical Newtonian physics may already have the answer and may be the key to the whole explanation is the speed of light.

Note[*]: for a full explanation of how the effects of GR explain SR refer to my book: '*The Enigma of Gravity.*'

In classical physics the speed of light is affected by the transparent medium through which it travels, in a vacuum the speed of light (c) is constant:

$$c = 300,000 \text{ kilometers per second}$$

Einstein said the speed of light *in a vacuum* is a constant, he did not say that the speed of light is constant, in fact he went out of his way to say that this constancy only applies to the empty vacuum of space free of gravitational fields. When light is travelling through the transparent medium of glass this speed drops to 200,000 km per second. In diamond, which is optically denser, it is even slower. It is the optical density of the medium which affects the speed of light, the more dense the medium the slower the speed of light. So this would suggest that in GTR, where a gravitational field slows down light, that the space vacuum is optically more dense. A gravitational field increases the optical density of the space vacuum (more imponderable matter per volume of space). Herein we have in GTR an explanation for the gravitational effects on the speed of light: the space vacuum is not an empty mathematical construct, but a transparent medium filled with imponderable (invisible) matter. The imponderable matter density (D) we will call mass-energy density. This leads us to two postulates:

1. The space vacuum imponderable matter (mass-energy) density (D) affects the speed of light.

2. The speed of light is inversely proportional to the space vacuum mass-energy density.

$$c = k/D$$

Where k is an exponential constant.

In this explanation for the variance of the speed of light for GTR we also have an explanation for time dilation: In defining time, only one explanation seems to make sense, we see the timeline for the evolution of the universe as ordered, timely, change. For the process of physical change, we need physical time, time can therefore be best defined as <u>rate of change</u> and herein the speed of light is the key; all change whether chemical or physical takes place over time and these chemical and physical changes are governed by the speed of light. All electro-magnetic interactions in chemical and physical reactions act at the speed of electromagnetic light. All electro-magnetic and gravitational forces (force fields) act at the speed of light, therefore we can state that:

Time is rate of change at the speed of light.

The speed of light governs time and change.

This explains time dilation in GTR:

An optically more dense region of space (a gravitational field) slows down light and therefore slows down physical and chemicals changes - slows down time.

Length contraction in a gravitational field can also be explained using this model:

In a gravitational field the optical medium of the space vacuum mass-energy density is higher (curved space-time) compared with normal space (flat space-time). The fact that this increasing energy density is increasing in the direction of a gravitating mass means that length in this direction is contracted.

Length in this medium of imponderable matter is contracted if its energy density increases - think of this as like having a one metre rubber ball ruler made with 10 rubber balls stuck together, squash the rubber balls closer together (more dense) and the ruler contracts in length.

Note: the odd thing about matter particles is that more massive particles are physically smaller in size.

This model of the space vacuum, that it is an optical medium filled with imponderable matter at various densities seems to explain all the effects of GTR.

The curvature of space-time being a smoothly varying space vacuum mass-energy density of imponderable matter

The speed of light would be measured as (c) in curved or flat space, in curved space since length *and* time are contracted *equally,* the measured value of the speed (c) would not change. Along a contracted ruler, light travels slower and so the measured speed would be the same. Even with the slowing down of time in a gravitational field an observer in this field would not notice clocks ticking slower because his brain and body and all physical processes (chemical and physicals reactions) slow down equally.

This model for the space vacuum, if true, should also explain the effects observed in STR.
In Einstein's Special Theory of Relativity, time slows down and mass increases exponentially as the speed of light is approached. Let us imagine a spacecraft travelling through this imponderable matter of space and gradually approaching the speed of light. What effects would this imponderable matter of space have on the spaceship and the environment and its crew inside?

Travelling at near light speeds means that the energy **density** inside the space ship has increased exponentially, the medium of space and its imponderable matter is flowing through the space ship at the same speed as the craft is moving through space. A beam of light trying to propagate inside the space ship would be slowed down, thus slowing down time, length would also be contacted since space-length has contracted (more dense) along the direction of motion, since the energy density is higher, mass would increase for there is more space vacuum drag on any relative acceleration.

Mass Dilation

It takes a Newtonian force to accelerate a mass through the vacuum of space and in Newton's second law of motion the amount of force needed to accelerate an object by a given amount is its measure of mass. This is what Einstein describes as inertial mass (m_i):

inertial mass = force applied/acceleration gained

But why does it require a force to accelerate a mass through the empty vacuum of space? Space appears to resist acceleration; it applies a drag force, similar to the drag of water when pushing an object through it, except that this drag only occurs during relative acceleration between the mass and the medium of space. The answer to this phenomenon of inertial mass lies in the fact that space is not empty but filled with 'imponderable matter' which resists acceleration.

Relative acceleration between a mass object and the imponderable matter of the space vacuum exerts a drag force on the object - action and reaction

Another measure of mass, is gravitational mass (m_g), we can measure mass by the amount of gravitational force acting on a mass - its weight. Again Newton gave us the answer:

gravitational mass = force experienced (weight)/gravitational acceleration

Einstein noted that there were two ways for measuring mass and that whichever way you used: the force applied to accelerate an object or the weight of an object in a gravity field; both would give you the same answer.

6. The Illusion of Mass

One of the major problems for interstellar space travel is mass. In order to place a spaceship in orbit about our earth requires enormous rocket propellant thrust to overcome the spacecraft's **gravitational** mass, then subsequent acceleration of the space ship to near light speeds also requires rocket propellant thrust to overcome **inertial** mass, but this is not the end of the problem, as the craft approaches near light speeds, its **relativistic** mass increases exponentially thus making it impossible to reach the speed of light. It seems, therefore, that the mass of the spacecraft presents us with a fundamental problem of reaching luminal and superluminal speeds.

What is mass and why does gravity, acceleration and near light speeds affect mass?

If we could eliminate these effects on mass we could easily reach star ship speeds for interstellar space travel.

The Illusion of Mass

1. **Inertial Mass**: caused by relative acceleration through space.

A mass stationary in space or in constant motion will only appear to have mass if we try and change its velocity (acceleration) relative to the space in which it is immersed. We 'feel' an object's inertial mass when we push it with a force or try and slow it down or change its direction.

2. **Gravitational Mass**: experienced as a force we call weight because of the **relative** acceleration of the space vacuum and a mass

3. **Relativistic Mass:** the increased force required to accelerate a mass through space as it approaches the speed of light.

In all cases mass is only experienced when we try to **relatively accelerate a mass through space**, something the Higgs field theorists neglect to state. It seems therefore that in order to build space craft which can travel at interstellar speeds we need to overcome the mass-effect of space on acceleration. This new science is called Quantum Engineering of the Space Vacuum.

The mass energy density of the space vacuum exerts a mass effect on an object during acceleration and limits the speed of light and therefore quantum engineering of this mass-energy density of the space vacuum seems the only way forward. The future of space travel lies in methods of slip streaming through the space vacuum where these relative acceleration mass effects are reduced.

This type of quantum engineering will in principle give rise to anti-gravity and space-time warp drives.

THE KEY

Since inertial mass is produced by relative acceleration of an object through the space vacuum, does this mean that gravitational mass is a relative acceleration of the space vacuum *through* the mass?

The answer to this must be yes, gravitational mass must be the drag force of the space vacuum imponderable mass-energy as it accelerates down through the object, thus pinning it to the floor.

7. THE KEY - Origins of Mass

The following discussion is quite 'lengthy' but it leads to some very important concepts concerning gravity.

Einstein was pleased to theorize that accelerated motion of a space craft through space produced 'artificial' gravity on board the craft. There was no difference in his mind, between the effects of acceleration and gravity, and therefore gravity was acceleration. In General Relativity gravitational acceleration was a space-time distortion created by a mass. One could not tell the difference between standing on a planet and feeling the effects of gravity or standing inside an accelerating space craft and feeling the same effects, one's own weight.

This must logically lead to the conclusion that when standing on a planet, something must be accelerating!

Since the two situations are equivalent, then it must be that when you are standing on a planet, space-time is accelerating through you.

This acceleration of the fabric of space through a stationary mass creates a force on the mass (you) which you feel as weight. Throw an object up in the air and it is accelerated back down by the physical acceleration of the space vacuum down into the Earth. This model would suggest that space is not empty, but must have substance. This physical substance of the space fabric Einstein referred to as 'imponderable matter' (ether).

In the left hand picture above the rocket is accelerating upwards through space and therefore space is accelerating downwards through the rocket and in the right hand picture, the same situation exists. In the second picture there must be a relative acceleration between the rocket and the space vacuum, and since the rocket is not moving relative to the Earth, then space must be the accelerating down through the rocket.

There remain two questions following on from this model of gravity

(In my book *The Enigma of Gravity*, I call this model of gravity Quantum Gravitational Flux Theory (QGFT):

1. If gravity is an acceleration of space into a mass and the accelerating space exerts a force on a mass placed in this flux, then space must have substance, so what is the physical nature of the substance of space?

Answer: Possibly an energetic flux of an exotic quantum soup of neutrinos, virtual (short lived) photons and virtual particle pairs.

2. If a mass such as the Earth causes an acceleration of this quantum soup into it, what property of matter particles (atoms) in the Earth or any other matter causes this acceleration of space?

Answer: matter particles interact with this energetic quantum soup by absorption and emission, this interaction being evidenced by known quantum energy shifts in sub-atomic particles.

Einstein refers to the mathematical geometry of space as space-time, the 3 dimensions of space and the 4th of time.

Distortion of this four dimensional geometry of space by a gravitating mass such as the Earth causes an acceleration of a test mass placed in this distortion.

But if space is being distorted, what exactly is being distorted? The General Relativity Theory of Einstein tells us that it is a distortion of the geometry (fabric) of space-time. In physics we can measure the dimensions of space (using a ruler) and the dimension of time (using a clock). Also in physics, anything which can be physically measured must have physical reality. In order to understand the physical nature of space-time we must treat space and time separately and then see how one is linked to the other and how a mass distorts both as space-time.

In his 1920 Leyden lecture Albert Einstein acknowledges that space must have substance.

In the 20th century the ether hypothesis became unfashionable since the substance of space, the ether, could not be detected by any experiments. Today we have two phenomena which prove the existence of the physical nature of the space vacuum:

1. **Lamb Shift** of the energy of electrons: light emission lines have 'fuzziness' to them. This is due to self-interaction of the electron and the electrons interaction with the space vacuum.

Most of the Lamb shift can be explained by two basic phenomena: self-energy and polarized vacuum:

a) **Self Energy:** the self-energy effect occurs when electrons subjected to electric and magnetic fields spontaneously and randomly emit photons and, within an incredibly short time span, reabsorb them.

b) **The polarized vacuum effect**: this involves pairs of virtual (short lived) particles one with a positive charge and one with a negative charge. Such pairs are created spontaneously from a vacuum in the presence of a positive or negative field and then are polarized, or aligned, by the field's charge.

2. **Casimir Effect**: two mirrors in a vacuum are attracted to each other. This results from a pressure difference in space, where it is greater outside the plates than between them. (Only certain modes of electromagnetic standing wave oscillations can exist between the plates (half wavelengths), whereas outside an infinite number of wavelengths are allowed.

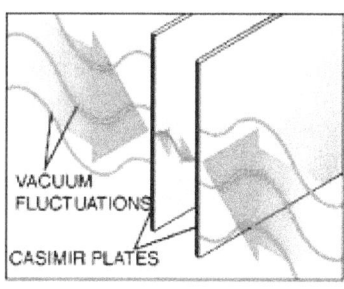

Summary
Quantum effects in atoms, energy shifts, are indicative that matter particles interact with the energy of the space vacuum as evidenced in Lamb Shift and the Casimir Effect. There may be a continuous flux of this energetic space vacuum into matter, which as a collapsing sphere of space around the Earth the acceleration rate (g) of which would follow an inverse square law.

Any mass particle placed in this energetic flow would experience a drag and be taken along with the flow, this we would see as free fall. A mass on the surface of the Earth would experience the drag of the space vacuum flux through it, this we would see as weight (W = mg). Here (m) equals gravitational mass and (g) the acceleration of space.

This is equivalent to inertial acceleration where by Newton's 2nd Law a mass particle accelerated through the space vacuum experiences a drag which we see as inertial mass. Since as stated above, any relative acceleration between the space vacuum and a mass causes a drag force on the mass, therefore gravitational mass and inertial mass are equivalent:

inertial mass = gravitational mss.

It is possible that this ethereal flow of space vacuum energy into matter not only causes matter particles to shift in energy levels as they absorb the energy quanta, but that the energy flux sustains matter in the ground state.

The following bits of mathematics are provided for evidence of the conclusions; you may skip the maths and go straight to the conclusions:

The ratio of the Lamb Shift energy (E_L) of a hydrogen electron to the mass-energy (mc^2) of the electron gives a value close to G:

$$\textbf{G approximates } E_L/mc^2$$

Where G is Newton's Universal Gravitational Constant which determines the magnitude of gravitational forces.

The exact value of G may be an average of all Lamb energy shifts in matter.

$$\textbf{Since } E_L = (alpha^5)(mc^2)k$$

Where alpha is the Fine Structure Constant and where k depends on the electron configuration and type of atom, then:

$$\textbf{G} = E_L/mc^2 = alpha^5 k$$

Conclusion
Thus G is proportional to the fundamental constant of electromagnetic interactions, alpha, the Fine Structure (God) Constant. This makes sense since alpha determines the strength of particle interactions with the quantum energy of the space vacuum.

Using Newton's universal Law of Gravitation:

Newton's gravitational constant (G) is a constant which determines the strength of a gravitational field. Calculating the base units for G using Newton's formula for gravity:

Since $F = GMm/r^2$

F = force between two masses in newtons (N)
M = mass one in kilograms (kg)
m = mass two in kilograms (kg)
s = time in seconds (s)
r = the distance between the centre's of the two masses in metres (m)

Since $F = GMm/r^2$

Then $G = Fr^2/Mm$

$= N.m.m/kg.kg$

Since $F = ma$

Then $N = kg.m/s^2$

Substituting the base units for newtons (N) we get:

$G = kg.m.m.m/kg.kg.s^2$

Thus $G = m^3/kg.s^2$

Since m^3 is **volume** and s^2 is **rate of change** then:

$G = $ Volume$/s^2$ per unit mass

Conclusion
Indicating that this gravitational constant is related to:

The rate of change of volume (of space) per unit mass

This demonstrates that base units for G are equivalent to the rate of change (flux) of volume (of space) per unit mass, suggesting that gravity may be an accelerating flux of the space vacuum energy into a mass.

If this model is correct then gravity is part of a great energy cycle in the universe.

Another way of looking at this mass equivalence is by the way we measure the acceleration experienced by inertial and gravitational mass:

Measuring Acceleration Using Inertial Mass

Accelerometers

An accelerometer is a sensor for testing the acceleration along a given axis. When a physical body accelerates in a certain direction, it becomes subject to a force equal to:

$$F = m_i$$

In accordance with Newton's Second Law.

In this formula, (m_i) is the mass and (**a**) the acceleration. Therefore, accelerometers are built on the principle of measuring the force exerted on a test body of a known mass along a given axis. The following drawing schematically shows the structure of an accelerometer.

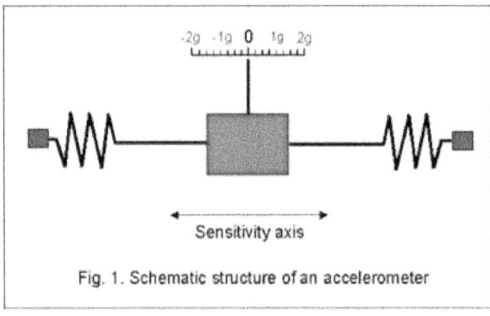

Fig. 1. Schematic structure of an accelerometer

In Newton's day, accelerometers where built using a test mass (shown in red) held at rest with springs and having a scale showing the acceleration along the sensitivity axis. Note that the unit (**g**) is equal to the acceleration subject to all bodies at the surface of the earth due to gravity, and is equal to about 9.8 meters per second per second. The same gravity is the acceleration that translates our body mass to a weight we can measure when we stand on a scale.

Linear Acceleration

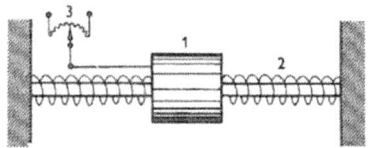

If the device is moving at constant velocity, the mass (1), supported on a bar by springs (2), remains static and an intermediate reading is registered on the zero centred voltmeter (3). This corresponds to zero relative acceleration between the mass and the space-time frame of reference. On acceleration in the direction of the supporting bar, i.e., along the accelerometer's sensitive axis, inertia causes the mass to lag behind, compressing the spring behind it and stretching the spring ahead of it: a positive voltage is registered. On deceleration, inertia causes the mass to compress the spring ahead of it and stretch that behind it, and thus a negative voltage is registered on the potentiometer. This corresponds to relative acceleration between the mass and the space-time reference in which it sits.

Stationary and Moving at a Constant Speed

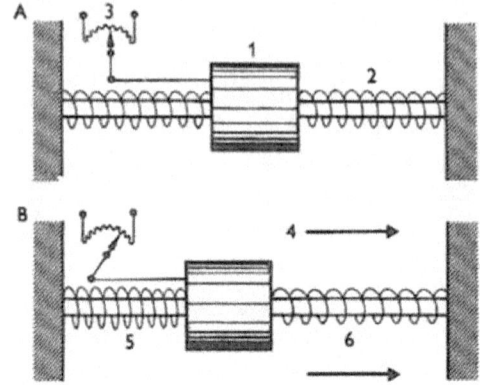

When the system is moving at a constant speed it gives the same result as a stationary system: the mass remains in the middle and at equilibrium with the springs ie *no relative acceleration.*

Conclusion

The spring system during acceleration exerts a force on the inertial mass, the mass (mass being its *resistance to acceleration*) exerts an equal and opposite force on the spring system. Newton's 3rd Law: to every action there is an equal and opposite reaction. So why is the mass reluctant to **relative acceleration** within the (**static**) space-time frame of reference? There must another 3rd Law pair of forces making the mass resist relative acceleration:

1st 3rd Law Pair of Forces:

Force of Spring on Mass = - Force of Mass on Spring

2nd 3rd Law Pair of Forces:

Force of Mass on Space-Time Reference Frame = Force* of Space-Time Reference Frame on Mass

*This is the resistive force *to relative acceleration between the mass and the space-time reference frame*, what I call **inertial drag** on the mass by space-time whenever there is relative acceleration between the two, this *drag force of space-time is what we see as mass.*

But how can space-time exert a drag force on a mass during acceleration? Einstein pointed the way to this in his mass equivalence theory.

2. **Gravitational mass** is a quantitative measure that is proportional to the magnitude of the gravitational force which is exerted by an object (active gravitational mass), or experienced by an object (passive gravitational force) when interacting with a second object.

Important Observation
What is interesting about an accelerometer is that if it is placed stationary on a table in the Earth's gravitational field, but orientated vertically it give this result:

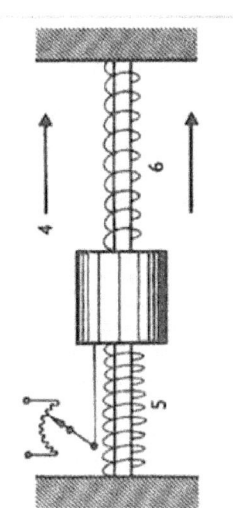

The mass moves downward from the equilibrium position.

This demonstrates that there is a relative acceleration between the system and the space-time matrix. Since the system is not moving upwards relative to the Earth or the observer, then the space time matrix in which it is immersed *must be accelerating downwards into the earth* thus pushing down with a drag force on the gravitational mass (m_g).

Here: **Weight = $m_g g$**

Where g is the gravitational acceleration of space.

Conclusion

Inertial mass and gravitational mass are equivalent, where *the mass is in relative acceleration with the space-time matrix:*

$$m_i = m_g$$

$$F/a = W/g$$

$$W = m_g g = F = m_i a$$

Where W and F are the equivalent drag forces which the mass experiences when in relative acceleration with the space-time matrix.

In simple terms, when holding out by the hand a mass and we experience the weight of the mass, we are actually experiencing the *drag of the space-time matrix which is accelerating into the earth* and when we try and throw a mass we also experience the drag of the space-time matrix, this is because in both cases there is a relative acceleration between the mass and the space-time matrix.

8. Understanding the Evidence for the Big Bang and the Age of Our Universe

Our understanding of the concept of time as being a constant is erroneous, true for an observer in his own space-time frame, but not relative to another observer in a different space-time frame. Time changes depending upon position (in space and in time) and speed and it is incorrect to extrapolate far back to the beginnings of our universe and state its age, for over that time-period space and time may have changed. It is enough to look back at the unfolding ages of the universe and state that it has undergone 13.8 billion <u>years of change</u>, but to state that it happed 13.8 billion years ago relative to the rate of modern time is incorrect.

With regard to the evidence for the expansion of the universe i.e. the red shift of light from distant galaxies, there are other explanations for this reddening of light over distances and time, one of which is:

Gravitational Red Shift

In Einstein's GTR light leaving a gravitational field into the flatness of space will increase in wavelength, become redder. The early expanding universe was hot, contracted and denser in terms of mass-energy, therefore light leaving the Big Bang gravity well to the present day expanded universe will experience gravitational red shift. When we observe distant galaxies we would expect to see their light red shifted even if they were now motionless and not expanding away from us. In observing more distant earlier galaxies we would expect to see more red shift.

This model would suggest expansion to a static universe. How would this gravity well affect time over the evolution of the universe? If the space vacuum was more dense with imponderable matter in an early more dense universe, then at that time, time would tick slower than it does today and in this early universe light and all electromagnetic reactions would also run slower, so looking back at these early processes we may see 13.8 billion years of change but the whole process may have taken an eternity to unfold and over time-change everything speeded up to its current rate of time-change.

The converse to this is even more interesting, assuming the early space vacuum was almost devoid of imponderable matter (not yet made by stellar processes), then time and change would have started at a phenomenal rate and gradually slowed to the current rate of time change. Looking back then to our remote past we may be observing 13.8 billion years of change, not time, and that it all came about in a matter of current earth time years or even days! The universe may only be several thousand 'years' old.

We cannot state that the universe is 13.8 billion years old, only that it has undergone 13.8 billion years of change.

Newton's Gravitational Constant in an Expanding Universe

In my book '*The Enigma of Gravity*' I pursue through classical and modern physics the cause and physical nature of gravity concluding that gravity is an acceleration of the imponderable substance of the space vacuum into a gravitating mass, thus gravity is an accelerating energy flow into matter, this invisible energy flow not only is it gravity but it may also sustain the very physical nature of matter and the universe. Gravity may be part of a universal energy cycle involving stellar, black hole and quasar processes, the space vacuum imponderable matter mass-energy density and ordinary matter. This invisible energy flow may even be responsible for stellar heating, and therefore solar nuclear fusion and planetary heating. There is only one thing in nature that can detect this flow, anything which has mass. Throw a ball up into the air and it will be dragged back down and pinned to the floor by this gravitational energy flow. Hold out you arm for some time and feel the weight (drag) of this flow. Earlier I showed that the derived units for Newton's Universal Gravity Constant G is related to acceleration of space volume into a gravitating mass:

$$G = m^3/kg.s^2$$
Since m^3 is volume and s^2 rate of change then:
$$G = Volume/s^2 \text{ per unit mass}$$
Indicating that this gravitational constant is related to:
The rate of change of volume (of space) per unit mass

This demonstrates that the base units for G are equivalent to the rate of change of volume (of space) per unit mass, suggesting that gravity may be an accelerating flux of the space vacuum energy into a mass.

The value of Newton's gravitational constant (G) determines the strength of gravity. It has a very small value; hence gravity is classed as a very weak force even though it is one of the most fundamental and most important forces in the universe:

1. Gravity puts the breaks on expansion, because gravity acts between every mass object in the universe, pulling them together.

2. Gravity is the force that holds galaxies and solar systems together.

3. Gravity drives the nuclear processes within stars to produce heat and light.

4. Gravity keeps the planets in orbit around a star such as our sun.

5. Gravity keeps us, our atmosphere, our oceans and everything else upon the earth.

6. Gravity caused the early universe primordial matter to clump together and form planets and stars.

7. Gravity, as an energy flow, may be the mechanism which holds and sustains the universe.

For all of the above processes the magnitude of the constant G is critical, for example too large and the universe would have expanded and collapsed under gravity very quickly. Too small and the universe would have expanded at an ever increasing rate. Its magnitude is just right for all of these processes to occur and stars, planets and even life to exist. So what determines the magnitude of the gravitational constant? Does the value of G change over time?

To answer these two important questions we need to understand what special physical property of mass particles (atoms) cause the acceleration of space-time. All mass objects, independent of the elements they are composed of, produce gravitational acceleration, so the cause of gravity is not related to the type of atoms present in a substance, but only to the amount of substance i.e. more mass (m) equals more gravitational acceleration (g).

So the cause must be related to the fundamental structure of the atom. Atoms are not passive structures; they are composed of dynamic particles, protons, neutrons and electrons. These particles are in constant oscillatory motion and are composed of electromagnetic energy (light) oscillating at frequency (f) at the speed of light (c). It is best to think of an atomic particle as a standing wave* of electromagnetic energy (E).

* When a guitar string is plucked its oscillation is a standing wave.

The mass-energy of a particle is given by Einstein's famous equation:

$$E = mc^2$$

So mass is equal to the amount of energy per speed of light squared:

$$m = E/c^2$$

Using Planck's constant (h) and his formula ($E = hf$) for the energy of a wave we can derive a new formula for mass:

$$m = hf/c^2$$

And since $c^2 = 1/(U_o E_o)$ where U_o is the magnetic constant of free-space and E_o is the electric constant, then:

$$\textbf{mass} = \textbf{hfU}_o\textbf{E}_o$$

This equation shows that mass is proportional to the frequency (f) of oscillation of an atomic particle <u>and</u> the electromagnetic properties of the space-vacuum.

Since gravitational acceleration is proportional to mass, it is fundamentally proportional to the frequency of mass-energy oscillation in an atomic particle. This would suggest that atomic particles are the dynamic machines which cause the acceleration of the space vacuum energy. If mass particles are absorbing this space vacuum energy (the imponderable matter), where does it go?

As surmised, this gravitational energy absorbed by matter may be sustaining matter, but what goes in must come out? There is evidence that oscillating atomic particles (namely the electron) interact with the space vacuum sea of energetic particles (the ether). This interaction causes the electron to jitter in energy levels, this jitter is called the Lamb Shift and analysis of the magnitude of this energy jitter is equivalent to <u>microwave radiation</u>.

A possible model for gravitational acceleration of the space vacuum sea of imponderable mass-energy is that:

Oscillating mass particles interact with the imponderable mass-energy of the space vacuum, thus absorbing energy out of the space vacuum causing its acceleration into a mass, the energy absorbed maintains matter (the particle wave standing oscillations) and in the process the space vacuum mass-energy is converted to microwave energy.

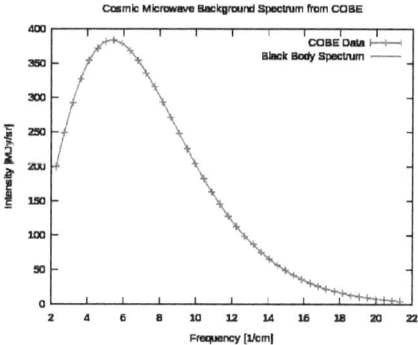

<u>If this model for gravity is correct</u> it would explain why Cosmic Microwave Background Radiation can be found in every direction in the universe a radio telescope is pointed. It could also explain planetary (microwave) heating and stellar heating which causes nuclear fusion in the core of the sun. Jupiter radiates more energy than it receives from the sun, the Earth has remained hot and molten for billions of years, and in both cases this related to the fact that the Earth (and maybe Jupiter) has a metal core: metal in a microwave oven heats up as it absorbs the microwave radiation.

Whatever the fundamental cause of gravity it may certainly not be a constant over the history of the universe, for if gravity is related and dependant on the imponderable substance of space (U_oE_o), then the energy density of the space vacuum will determine (fix the magnitude of) Newton's gravitational constant.

If the energy density of the space vacuum decreases with time as the universe expands, then G will become smaller and thus gravity weaker, this may be the cause of the recently discovered accelerated expansion of the universe.

9. Borrowed Energy

There are many accounts of the origins of all things, myths, legends, stories, beliefs and of course scientific theory, but one of the questions that my physics students ask me concerning the Big Bang is that how can something come out of nothing, if there was nothing before the Big Bang where did all this (the universe) come from? From the Biblical Genesis account, what was before creation was God, but leaving beliefs aside science has some answers:

Firstly, the question of 'before the Big Bang,' since time and space are inseparable, without space and matter there was no time, there was no before. 'Outside of time, matter and space,' time has no meaning. There just is. It is the physical nature of this universe which limits the speed of light and therefore limits the speed of change, outside of this physical universe light has no limit and time has no meaning.

The universe is made up of mass-energy, matter and light which occupy space, without this universal substance, space would contract to nothing, for space itself (the 3 dimensions of volume) is a physical mass-energy substance, the imponderable matter. So where did all this mass energy come from?

If I throw a ball up into the air, the ball has been given mass-energy (kinetic energy) and as it rises this kinetic energy decreases, the ball slows down to a stop at its highest point and all of its original kinetic energy is converted to Gravitational Potential Energy (GPE) stored within the surrounding gravitational field. This is a bit like stretching an elastic band where the original kinetic energy to stretch the band is stored in the elastic (strain energy), the bonds between the molecules are stretched. The ball returns to my hand during which the GPE is converted back to kinetic energy and this energy is given up by the ball back to the hand which initially threw it up.

During the whole time of its rise and fall the ball had no energy of its own, it was borrowed energy and also during its rise and fall its total energy was a constant, for at every stage the sum of the potential and kinetic energies is constant.

This is like our own universe the kinetic energy of the Big Bang has been converted to the mass-energy of particles and gravitational potential energy due to the separation of masses against the force of gravity, so likewise the universe has no energy of its own, it is borrowed energy and at all stages of its expansion its total energy is constant. Where was this energy borrowed from? Science has no answer to this question, but in religious beliefs the mass-energy was borrowed from the hand of God.

10. Through the Looking Glass - Our View of the Universe

'If you don't know where you are going, any road will get you there.'
(Lewis Carroll)

We are time frozen in a universe which has only one window, the heavens, and like Alice's mirror it shows us wonderful things. Our ancestors were limited by the view gained by naked eye observations, with the invention of the telescope we have been able to see deeper and deeper into the heavens, back to a time unimaginable and we see wondrous things. The earliest stars and galaxies formed only millions of years after the initial Big Bang. Our telescopes not only allow us to see deeper into space, but further back in time. The latest images from the Hubble Telescope allows to view back into the past some 13 billion years, when the universe was only 0.7 billion years old.

Evidence of Accelerated Expansion and the Early Hot Universe

The assumption: Since light takes one year to travel a distance of one light year these galaxies are at a distance of 13 billion light years.

The problem: Because the universe is expanding, the question of the distance to a very distant galaxy is hard to answer. It all depends on your point of view. This is the problem of defining a distance in an expanding universe:

Two galaxies which are near to each other when the universe was only 1 billion years old and the first galaxy emits a pulse of light. The second galaxy does not receive the pulse until the universe is 14 billion years old.

By this time, the galaxies are separated by about 26 billion light years; the pulse of light has been travelling for 13 billion years; and the view the people receive in the second galaxy is an image of the first galaxy when it was only 1 billion years old and when it was only about 2 billion light years away.

There are four different distance scales commonly found in cosmology:

(1) Luminosity Distance - DL: In an expanding universe, distant galaxies are much dimmer than you would normally expect because the photons of light become stretched and spread out over a wide area. This is why enormous telescopes are required to see very distant galaxies. The most distant galaxies visible with the Hubble Space Telescope are so dim that they appear as if they are about 350 billion light years away even though they are much closer.   Luminosity Distance is not a realistic distance scale but it is useful for determining how faint very distant galaxies appear to us.

(2) Angular Diameter Distance - DA: In an expanding universe, we see the galaxies near the edge of the visible universe when they were very young nearly 14 billion years ago because it has taken the light nearly 14 billion years to reach us. However, the galaxies were not only young but they were also at that time much closer to us. The faintest galaxies visible with the Hubble Space Telescope were only a few billion light years from us when they emitted their light. This means that very distant galaxies look much larger than you would normally expect as if they were only about 2 or 3 billion light years from us (although they are also very, very faint). Angular Diameter Distance is a good indication (especially in a flat universe like ours) of how near the galaxy was to us when it emitted the light that we now see.

(3) Comoving Distance - DC: The Comoving Distance is the distance scale that expands with the universe. It tells us where the galaxies are now even though our view of the distant universe is when it was much younger and smaller. On this scale the very edge of the visible universe is now about 47 billion light years from us although the most distant galaxies visible in the Hubble Space Telescope will now be about 32 billion light years from us. Comoving Distance is the opposite of the Angular Diameter Distance: it tells us where galaxies are now rather than where they were when they emitted the light that we now see.

(4) Light Travel Time Distance - DT: The Light Travel Time Distance represents the time taken for the light from distant galaxies to reach us. This is what is meant when it is said that the visible universe has a radius of 14 billion light years - it is simply a statement that the universe is about 14 billion years old and the light from more distant sources has not had time to reach us. Light Travel Time Distance is as much a measure of time as a measure of distance. It is useful mainly because it tells us how old the view of the galaxy is that we are seeing.

For small distances (below about 2 billion light years) all four distance scales converge and become the same, so it is much easier to define distances to galaxies in the local universe around us.

Redshift is a measure of the stretching of light caused by the expansion of the universe: a galaxy with a large redshift is further away than a galaxy with a small redshift. The most distant galaxies visible with the Hubble Space telescope are at redshift 10, whereas the most distant protogalaxies in the universe are probably at about redshift 15. The edge of the visible universe is at redshift infinity. A typical portable telescope, by contrast, cannot see very much beyond redshift 0.1 (about 1.3 billion light years).

The Luminosity Distance (DL) shows why distant galaxies are so hard to see: a very young and distant galaxy at redshift 15 would appear to be about 560 billion light years from us although the Angular Diameter Distance (DA) suggests that it was actually about 2.2 billion light years from us when it emitted the light that we now see. The Light Travel Time Distance (DT) tells us that the light from this galaxy has travelled for 13.6 billion years between the time that the light was emitted and today. The Comoving Distance (DC) tells us that this same galaxy today, if we could see it, would be about 35 billion light years from us.

Our looking glass through which we view the heavens has become more and more sophisticated, but it is in the interpretation of what we see which is of the most fundamental importance and until we can journey to these distant stars and galaxies and observe at first hand we will always remain seeing our past 'through the looking glass.'

11. The Big Bang - In a State

Extrapolation of the expansion of the Universe backwards in time yields an infinite density and temperature at a finite time in the past. This, what is called a singularity, signals the breakdown of general relativity. How closely we can extrapolate towards the singularity is debated. This singularity is sometimes called the Big Bang, but the term can also refer to the early hot, dense phase itself, which can be considered the 'birth' of our Universe. Based on measurements of the expansion using Type Ia supernovae, measurements of temperature fluctuations in the cosmic microwave background, and measurements of the correlation function of galaxies, the Universe has a calculated age of 13.75 billion years.

In a State - The Phases of Expansion

It is better to look at this process in reverse, starting with the matter filled universe at a temperature just above absolute zero (2.725 kelvin) and contract the universe to a point of extremely high density and temperature. We are all familiar with the three states of matter: solid, liquid and gas. We understand that to change a solid (ice) to a liquid (water) requires heat energy; the same is true when there is a change of state from a liquid (water) to a gas (steam, more correctly, water vapour.)

It takes energy to separate the atoms for each change of state, in a solid the atoms are strongly bonded together and very close, in a gas there are very weak bonds between the atoms and they whiz around at high speeds with much greater separation distances. What happens if we continue heating a sample of matter such as hydrogen?

There is a fourth change of state and the hydrogen gas changes into hydrogen plasma, during this change of state the negative electrons are ripped away from the positive nuclei (single protons). The hydrogen plasma is a hot soup of electrons and protons; we have broken down matter into its sub-atomic particles. Our plasma soup also contains heat radiation, light photons. The sun is so hot that most of its hydrogen is in a plasma state (6000 - 13,000,000 degrees Celsius). What happens if we continue heating and raise the temperature further? Is there a fifth state of matter?

Since the matter in the universe is mostly hydrogen we will look at what happens to hydrogen as the universe gets hotter and hotter:

Theoretically if we continue to heat hydrogen plasma it will undergo another change of state:

The fifth state of matter theoretically will be a quark, electron and photon soup. The protons breakdown into sub-atomic particles called quarks. This happens around 10^{15} kelvin.

If we keep on heating then we will enter another phase state where heavy particles appear, called bosons, so we now have the sixth state of matter: quarks, electrons, bosons and photons.

There is a seventh state of matter, this is when all the particle constituents, quarks, electrons and bosons break down into pure energy light photons, yes the seventh state of matter is pure light energy. We are now between 10^{27} and 10^{32} kelvin and a time of 10^{-34} to 10^{-44} seconds in the life of the universe, just a tiny fraction of a second into the initial Big Bang (Time zero).

12. Truth is not Subjective

There is a fundamental question that has been asked by many generations, 'where did all this (the universe) come from?' Historically, this desire to know the truth concerning the origins of life and the universe, may only be a modern enigma. Our ancestors had no issue with the existence of the universe, or how and why it came into being; it was simply an act of God or a god, which many cultures describe in a multitude of ways. There are some striking similarities in the creation stories and some diverse contradictions and differences. If one thinks about it in a simple sense:

1. The Universe Exists.

2. Matter and Space Exist.

3. Time Exists.

4. Light and Energy Exist.

5. Atoms and Subatomic Particles Exist.

6. The Forces Governing Matter, Space-Time, Atomic Particles and Chemical Interactions Exist.

7. The Physical and Mathematical Laws of the Universe Exist.

8. The Stars, Planets and the Earth Exists.

9. The Potential and Reality for 'Simple' Atomic Particles to Form Complex Organic and Inorganic Chemicals Exist.

10. The Potential and Reality for Complex Organic Chemicals Assembled in Even More Complex Structures and Metabolic Pathways to Form Fully Functional Reproducing Harmonic Units Exist.

11. Life Exists.

12. Sentient Life Exists, 'I think, therefore, I am.'

And so we are left with the classical questions: where, what, when, how, why and who? It is interesting to note that when it comes to the science of the origins of the universe and life itself, the fundamental questions of 'who?' and mostly 'why' are not confronted by science, this we are led to believe is the responsibility of religion or faith.

Let us start with the 'easy' ones first 'where and when?'

Where and when did all of what comprises the known universe come into existence? When we ask this question we are immediately faced with the obvious: the universe had a beginning and therefore must have an end and we are somewhere along a time line in what we call the present. This may be a wrong question to ask, the concept of 'when' involves time and time is part of the structure of this universe, without the universe, there is no time. Likewise, the question 'where' is meaningless for without the fabric of space-time, there is no 'where.'

The two questions 'who and why,' ascribes intelligence 'before' the universe came into existence, a creator God. This book only concerns itself mostly with the question, the scientific 'how?' This question can be treated with evidence and observation, logic and reasoning. It is in exploring the facts about this universe which may in its treatment reveal answers for the other fundamental questions.

The universe simply exists, it has no beginning and no end, and it is truly infinite in size and time. This does not mean that the processes within the universe did not have a beginning, time is change and the universe is and always has been in a constant state of change. The current state of the universe we will call the present and the earliest conceivable processes of change we will simply call the beginning of the universe. In examining these process of change (history) we are assuming that the rate of change (time) is constant; this may not be true as we have seen earlier. Let us start with the current observable state of this universe and then by looking at the evidence retrace its history.

The Current Observable State of the Universe

I am always amazed by the comments made by scientists and documentary presenters when they make the statement that we exist on an average planet orbiting an ordinary average star. There is nothing average or ordinary about our existence, planet Earth is a remarkable planet, and one which has the right conditions to support a remarkable ecosystem and our star, the sun is certainly not average or ordinary. There are many types of stars and most of them are not suitable for the support of life. They can be too small and too cold or too big and too hot with intense radiation that would immediately frazzle any life forms which dared to come into existence, these large stars also have very short lifetimes and are therefore not much use for habitable life on an orbiting planet.

Our star is an ideal size, age, temperature and composition to support life. If we ignore the percentage of hydrogen in the sun the remaining elemental percentage composition is the same as Earth's which means it is a kind of partnership, the Earth and the sun came from the same primordial matter and ended up together with this 'symbiotic' existence for the support of habitable life. Another remarkable thing about our planet is that it contains an abundant amount of the chemical needed for many chemical and biological reactions and for life: water. During the formation of the hot Earth, primordial water should have evaporated away into space, suggesting that Earth obtained its vast supply after its formation. There is no reliable evidence thus far that life exists elsewhere in the universe other than on Earth in the Sol System of our Milky Way galaxy. We are unique, the life forms on this planet are unique and the Earth-Sun system is unique. Other planets around other stars are currently being indirectly discovered by NASA's Kepler Mission. The Kepler telescope looks for minute changes in a star's luminosity during the transition of an orbiting planet. A transit occurs each time a planet crosses the line-of-sight between the planet's parent star that it is orbiting and the observer. When this happens, the planet blocks some of the light from its star, resulting in a periodic dimming. This periodic signature is used to detect the planet and to determine its size and its orbit. It may well be that one day another Earth like planet will be discovered which supports primitive life and complex sentient life forms, but until there is evidence to the contrary we must assume that life on Earth is unique.

The Apollo astronauts when looking back at our fragile beautiful blue water planet were moved by its very existence, Mother Earth may support the only life in the universe. If this is truly the case, then every individual on this planet has a unique existence.

In the future we may discover other life forms on other worlds, but until that time comes we have to acknowledge that we are a very important part of the existence of this universe.

The Beginning of Time

In physics one of the hardest questions to answer is: how, in the beginning, did the universe come into existence? Here science struggles with concepts that were once previously within the realms of philosophy and religion. The universe exists and we exist and therefore there must be an answer to such a profound question, if you believe in a Creator God, then this is not a problem, but science deals with facts and evidence and without a Creator God science has to come up with a rational answer, otherwise it has to accept the existence of God.

Therefore for an atheistic scientist there has to be an answer and that answer can be arrived at through logic, reason and evidence, but the same is true for a theist, if God created the universe then the evidence of the 'how' should still be apparent. In seeking the truth of how all this came into being, whether theist or atheist, by examining all of the scientific evidence (that which is discovered and that which is to be discovered), in time both groups should arrive at the same answer, the same truth. Cause and effect is the key to finding solutions to this type of problem. In forensic science a crime scene is littered with evidence as to how the crime was committed, of course the criminal investigator is also interested in the 'why, motive' and the 'who?'

One can imagine early primitive man looking up at the heavens and wondering about the nature of the heavenly lights, the endlessness of the ground on which he stood and his very existence, it is no wonder therefore that different groups, races and cultures came up with their own ideas to answer these questions and naturally a superior being was involved for to bring into existence such a remarkable world was certainly outside the power of mortal man. The ancient Greek legend of creation is quite poetic and imaginative:

The Greek Legend of Creation

In the beginning, Heaven and Earth existed, there were no gods. Heaven and Earth were the first parents and from their union sprang the Gigantic Titans. For ages the Titans ruled the Earth, but at last the gods, who were the children of the Titans, rebelled and overthrew them. Zeus became the supreme ruling god and his wife and sister, Hera became queen of heaven.

Now as yet there were no men on earth, and none of the animals seemed worthy to rule the rest. So the gods decided to make still another kind of creature. One of the Titans, Prometheus, whose name means Forethought, was chosen for the task.

Down from heaven the Titan sped. He took clay and mixed it with water, kneaded it and shaped it into the likeness of the gods. He made his creature stand upright, for he wanted man to look up at the stars and not down on the earth, like the animals. Then Prometheus thought: 'What gifts shall I give this work of my hands to make him superior to the rest of creation?'

Unfortunately, his brother Epimetheus, which means After-thought, had already given all the great gifts to the animals. Strength and courage, cunning and speed, he had distributed them to all. Wings, claws, horns, scales, shell covering, nothing was left for man. Then quick witted Prometheus thought of fire. Oh, great and wonderful gift! 'With fire,' the Titan thought, 'man can make weapons and subdue the beasts, forge tools, plough the earth, and master the arts. What matters that my creature has neither fur nor feathers, scales or shell? Fire will warm his dwelling, and he need not fear neither rain nor snow nor the wild north wind.'

It must be pointed out here that the ancient Greeks came late into humankind's history and had probably taken many ideas concerning creation from other cultures, the ancient Egyptians and biblical writers used the concept that man was made from clay, mud and water, and what is even more remarkable is that in modern science, the very first life forms are reckoned to come from the primordial mud pools of primitive earth, from simple organic chemicals combining into more complex organic chemicals until eventually forming simple, self replicating, single celled units.

'Truth is not subjective, neither is it objective,
it is simply truth.'

13. Cause and Effect

That which brought this universe into existence, cause and effect, is one of the biggest headaches for the modern physicist, the essence of the conundrum: physical effect follows (involving time) physical cause.

Before this universe physically came into existence there was no physical anything, including time, for effect to follow cause. How can something come out of nothing with no cause? This question assumes that that which exists and which is independent of physical space and time follows the same physical rules which exist in our universe, but it may not have any rules at all. It is sufficient to say that because this universe exists and evolved into what we see today there must be involved a process which left a fingerprint which can be examined by observation and evidence. Therefore it is the physical fingerprint of the origins of our universe which we shall examine in the light of that which we know and understand.

The evidence that our universe is expanding and cooling suggests that in the past it was hotter and contracted, extrapolation, a dangerous thing to do with any scientific evidence, suggests that the initial energy for hot expansion came from a Big Bang.

This leaves us with the following questions:

1. What is the evidence that the universe is expanding? The evidence for this comes from the Red Shift of light from distant galaxies, but there are alternative explanations that do not require expansion.

2. Where did the energy come from for the universe to explode into expanding matter and space-time?

Something Out of Nothing

In examining physical scientific evidence, one of the main concerns I have is that we can only observe and record that which we can physically measure. There may be 'forces' at work which on the purely physical level we will never be able to observe and easure, let alone analyze. Nevertheless, since we are limited by physical measurement, what exactly can we measure?

Fundamentally there are only three things we can measure:

length, mass and time.

The poor physicist has therefore very limited tools to explain any observable, physical effect and cause. These three physical parameters may seem limited, but in fact they are very powerful, even though they are limited to the physical realm.

It seems then since we can only measure that which is physical, that our theory of how the universe came into existence ignores one of the fundamental laws of science, cause and effect.

14. Something for Nothing

The fundamental properties of our universe are so entwined and interdependent that if we examine one of these properties without due regard to the others it will lead us to less of an understanding of their true individual nature.

The first three I have mentioned before: **mass, length and time**.

In this discussion when referring to mass we include Einstein's mass-energy equivalence. Mass we must understand is a form of locked up energy. There is a fourth physical phenomenon which lies at the heart of the first three and which binds them together in perfect symmetry: light.

The dimensions of length give us the three dimensional space in which we and all things exist and we would be foolish to think that space has no physical properties, the physical dimensions of space gives space length and matter position. Moving through the dimensions of space requires time and these three are summed up in the high school equation:

Speed (the rate at which we traverse space length) = length/time

Light traverses space at a fixed, constant, speed, the speed of light. This fundamental limit to speed reflects the intrinsic properties of space:

space limits and fixes the speed of light.

All physical communication of effects, whether by forces, chemical and atomic interactions etc occur at this speed, the speed of light. Since change is our measure of time, then:

the speed of light governs time.

Space limits the speed of light and light governs the rate at which time 'ticks.'

If we were to imagine a portion of space enclosed within a box and using the most powerful vacuum pump extracted every single atomic particle from within we would end up with what we may call a true vacuum. This is not the end of the story for since the box has temperature, heat energy, inside there will be photons of light (infra-red heat waves) bouncing back and forth within the box. In order to end up with a true vacuum devoid of any energy at all we would have to place our box inside a super fridge and reduce the temperature of the box to absolute zero (-273 Celsius, 0 kelvin). Now we have 'nothing' inside the box except the three dimensions of space.

But if I were to pass a beam of light through the box while inside the cold vacuum of space the light photons would still be traveling at the speed of light. Why would nothing affect something?

The answer must be that the box (space) is not empty at all and that empty space has physical properties which limit the speed of light.

It makes no sense thinking that empty space has normal fathomable physical properties in the way we understand them. The vacuum of space has physical existence but its properties are a little on the weird side. Within the box there still exists position, length, speed and time; it is just devoid of mass particles and radiant energy. Any mass particle put in motion inside this box would still obey Newton's laws of motion. The particle can travel at constant speed inside without any applied force and to change the motion of the particle, speed and/or direction would require a force. The empty vacuum of space enforces these laws, you can move through it freely, but to change your motion (acceleration) will cost.

So what is inside the 'empty' box?

an 'infinite' amount of mass-energy.

The 'imponderable' matter which occupies all space.

The space-vacuum can be thought of as being filled with some kind of super-fluid under enormous tension and pressure. The question of what type of exotic matter it is composed of still remains a mystery. Many scientists argue that this ethereal material does not exist, for it has not been detected: while others, although few in number, state that space-time cannot simply be a mathematical construct but since it has physical properties it must have physical substance, as stated by Einstein in his 1927 Leyden lecture. So what are the properties of this 'imponderable matter?'

1. Possesses length in three dimensions.

2. Limits the speed of light to a constant velocity (c)

3. Enforces time on physical, electro-magnetic, processes (rate of change).

4. Has a varying optical (mass-energy) density which is seen as the curvature of space-time, what we call a gravitational field.

5. Its increased optical density in a gravitational field slows down the speed of light (Shapiro Delay) and curves its path[*] and contracts length.

[*] **Note**: the fact that a gravitational field can curve (warp) the path of light in classical physics would be understood as refraction. Refraction of light, a bending towards the gravitating mass, would be accepted as evidence that the optical density of space increases towards the gravitating mass.

6. When there is relative acceleration between a mass-object and the space-vacuum, the space vacuum exerts a drag force on the mass-object which we interpret as inertial or gravitational mass.

7. The magnitude of the mass-energy density of the space vacuum and the rate of its acceleration into a gravitating mass as a collapsing volume sphere determines the magnitude of Newton's Gravitational Constant (G) and hence the magnitude of the acceleration of gravity (g) for any mass-object placed in this collapsing sphere of imponderable matter.

If this model of the ether is correct, then the question remains as to the nature of the imponderable matter. The refraction of light by a gravitational field which causes light bending and space and time dilation would infer that space is endowed with optical density of a particulate nature. In other words, the very fabric of space is made up of a soup of exotic 'invisible' atomic particles. Whatever particles make up this exotic soup, they have a very, very weak interaction with matter, as is the case for gravity. This would suggest that these atomic particles of the space vacuum are almost electrically neutral, have an almost imperceptible mass and size. There is one known contender that could fulfill this role, the elusive **neutrino**.

Neutrino's are a class of subatomic particles with zero electric charge and almost zero mass.

15. The Elusive Neutrino

Neutrinos are ghostly subatomic particles that, by all reasoning, should have a minute mass, but until now that mass has never been accurately calculated, which has posed problems with what's known as the standard model of cosmology. Neutrinos are difficult to study because they interact so weakly with matter.

Neutrinos are known to have different 'flavors.' While originally thought to be massless, these various neutrino flavors have recently been estimated to have a total mass of 0.06 electron volts (eV), which is far less than a billionth of the mass of a proton. A recent researcher report, suggests that the sum of the masses of three known neutrino flavors is 0.320 +/- 0.081 eV.

Using the latter figure of 0.320 eV this is equivalent to a mass of 0.57×10^{-36} kg, three billionths the mass of a proton. Using Einstein's mass-energy equivalence formula:

$$E = mc^2$$

Energy = mass x speed of light squared

Gives us the energy equivalent of the neutrino as 5.122×10^{-20} Joules

This is equivalent to a maximum particle wavelength (hence an estimate of its maximum size) of: 3.88×10^{-6} metres. Considering an atom is of the order of 10^{-10} metres, this makes the elusive neutrino:

**8000 times bigger than an atom
and three billion times less massive**

A quantum soup of electrically neutral neutrinos of large size and little mass would be almost completely transparent to real matter. An exotic soup of neutrinos may fulfill the properties needed for Einstein's 'imponderable matter.' Neutrinos rarely interact with matter and therefore normal matter is almost transparent to their motion. These properties are similar to gravity which acts *through matter* and is a very weak force *on* matter.

16. The Formation of the Earth

According to the Standard Model the Earth formed 4.5 billion years ago:

The standard model for the formation of the Solar System (including the Earth) is the solar nebula hypothesis. In this model, the Solar system formed from a large, rotating cloud of interstellar dust and gas called the solar nebula. It was composed of hydrogen and helium created shortly after the Big Bang 13.8 billion years ago and heavier elements ejected by earlier exploding star, supernovae.

About 4.5 billion years ago, the nebula began a contraction under gravity. As the cloud began to accelerate, its angular momentum, gravity and inertia flattened it into a proto-planetary disk perpendicular to its axis of rotation. Due to collisions of aggregating matter kilometer-sized proto-planets began to form.

The center of the nebula collapsed rapidly, the compression heating it until nuclear fusion of hydrogen into helium began. After more contraction, a star ignited and evolved into the Sun. Meanwhile, in the outer part of the nebula, gravity caused matter to condense and the rest of the proto-planetary disk began separating into rings.

In a process known as accretion, successively larger fragments of dust and debris clumped together to form planets. Earth formed in this manner about 4.54 billion years ago.

The proto-Earth grew by accretion until its interior was hot enough to melt the heavy metals (mostly nickel-iron). Having higher densities than the rock forming silicates, these metals sank. This resulted in the separation of a primitive mantle and a (metallic) core only 10 million years after the Earth began to form, producing the layered structure of Earth and setting up the formation of Earth's magnetic field.

During this time as the outer silica layers solidified into the solid crust the highly active volcanic Earth would have been continually bombarded by meteorites, the material left over from the aggregation of planets. This early hot Earth could not have sustained liquid water (or oceans) which boils at a temperature of only 100° Celsius. Volcanic out-gassing of water vapour would have continued as the Earth's exterior cooled down.

The Earth's Oceans and Atmosphere

The first atmosphere, captured from the solar nebula, was composed of light elements, mostly hydrogen and helium. A combination of the solar wind and Earth's heat would have driven off this atmosphere. The early molten Earth released volatile gases; and later more gases were released by volcanoes, completing a second atmosphere rich in greenhouse gases, nitrogen and water vapour but poor in oxygen. Finally, the third atmosphere, rich in oxygen, emerged when bacteria began to produce oxygen about 2.8 billion years ago.

At this time between the formation of the second and third atmosphere the Earth would not had enough water to form the oceans that we see today. It is supposed that early oceans formed from the arrival of extraterrestrial water arriving via icy comets and meteorites from the remaining nebula debris. As the planet cooled, clouds formed and rain created the oceans, but the origins of the vast supply of water on Earth, or the reason that there is clearly more liquid water on the Earth than on the other rocky planets of the Solar System, is not completely understood and remains a mystery. There exist numerous more or less mutually compatible hypotheses as to how water may have accumulated on the earth's surface over the past 4.6 billion years in sufficient quantity to form oceans.

So where did all Earth's water come from?

17. Water Earth

The theory that Earth's water originated purely from comets is implausible, as a result of measurements of the isotope ratios of hydrogen in four main comets, for according to this research the ratio of deuterium to protium of the comets is approximately double that of oceanic water. Other sources may have contributed to Earth's water:

1. **Hydrate Minerals**: gradual leakage of water stored in Hydrate minerals of the Earth's rocks.

2. **Volcanic Activity**: water may also have come from volcanism: water vapour originating in volcanic eruptions condensing and forming rain.

Either way, for a long time the jury was still out on identifying the source of Earth's vast water supply, that is until now. Latest geological evidence supports the fact that Earth's vast water supply may have been trapped within the Earth's mantle from the beginning and is a natural part of the Earth's geological history.

The Water Below

After decades of searching scientists have discovered that a vast reservoir of water, enough to fill the Earth's oceans three times over may be trapped hundreds of miles beneath the surface, potentially transforming our understanding of how the planet was formed.

According to the latest research the water is locked up in a mineral called ringwoodite about 660km beneath the crust of the Earth. The discovery suggested Earth's water may have come from within, driven to the surface by geological activity, rather than being deposited by icy comets hitting the forming planet as held by the prevailing theories.

Geological processes on the Earth's surface, such as earthquakes or erupting volcanoes, are an expression of what is going on inside the Earth, out of our sight. This is evidence for a whole-Earth water cycle, which may help explain the vast amount of liquid water on the surface of our habitable planet. Scientists have been looking for this missing deep water for decades. The research provides direct evidence that there may be water in an area of the Earth's mantle known as the transition zone.

Ringwoodite acts like a sponge due to a crystal structure that makes it attract hydrogen and trap water. If just 1% of the weight of mantle rock located in the transition zone was water it would be equivalent to <u>nearly three times the amount of water in our oceans</u>.

The research produced evidence that melting and movement of rock in the transition zone hundreds of kilometres down, between the upper and lower mantles led to a process where water could become fused and trapped in the rock.

One researcher reported that the hidden water might also act as a buffer for the oceans on the surface, explaining why they have stayed the same size for millions of years:

'If the stored water wasn't there, it would be on the surface of the Earth, and mountain tops would be the only land poking out.'

18. A Geological Inevitability

As discussed earlier it makes complete sense that the primordial water of the original solar nebular which formed the planets and our Earth finally ended up chemically trapped within the mantle. Vast amounts of water locked in mantle minerals are a natural part of our planet's geology. Lava flow convection currents moving down into the hotter regions of the lower mantle from the water bearing transition zone releases this water up through the cooler upper mantle and in to the Earth's crust for venting into the atmosphere via volcanoes. This forms part of a natural water cycle, were subduction of the oceanic crust under the thicker continental crust takes water and rock back down into the mantle melt.

The Inevitable Process

1. Originally, our molten Earth with its locked in primeval water cooled to a hot smooth sphere of solid crust floating upon a sea of molten lava.

2. Gradually the heavier metals (nickel-iron) sank to form the magnetic core and hence the earth's magnetic field.

3. The mantle below the global crust of probably equal thickness separated by density out into three zones:

a) **Cooler Upper Mantle**: where pressure and rising convection currents pierced the crust to form water venting volcanoes.

b) **Water Bearing Transition Zone:** where water locked up in mineral form is released by sinking convection currents into the upper mantle and accumulates beneath the thin global primeval crust.

c) **Hotter Lower Mantle**: where downward convection currents from the transition zone releases water for volcanic venting.

4. As the Earth continued to cool accumulated water vapour in the Earth's atmosphere formed a tropical type of environment of high humidity and temperature, where some of the atmosphere's super saturated water content condensed to form shallow seas. These hot primeval seas would have slowly increased in size and cooled along with the rest of the Earth. The increasing weight of these early hot seas pushed down on the primitive solid crust below forming deeper, but still shallow oceans while simultaneously pushing up low mountain ranges. During this time of the formation of early shallow seas and low mountain ranges the Earth's crust was intact and of equal global thickness.

This thin primeval crust has <u>been preserved</u> in what is seen today as the oceanic crust, which is much thinner than our ancient continental crust which has acquired more mass and thickness over time.

5. The geological processes occurring in (3b) would inevitably lead in time to a major global catastrophic event.

19. The Day the Earth was Smashed like a Cup

Our early hot tropical Earth of a water saturated atmosphere and hot shallow seas gradually cooled to an ambient environment suitable for habitable life. Jumping forward in time to when our early Earth was teeming with plant and animal life and early man, we would arrive at a place of low lying land masses covered in lush tropical forests surrounded by warm shallow saltier seas, a paradise of animal and plant variety. During this time of tropical paradise everything would have been watered by hot steaming vents (geysers) and active volcanoes. The atmosphere would have been saturated with the continuing out gushing of the hot primeval water from below, no need or opportunity for the cooler processes of cloud formation and rain. The variety of animal and plant life forms would have been a wonder to behold; giant reptiles roamed the land, the dinosaurs, which in later human folklore and myth would have been remembered as monsters, beasts and dragons.

Unfortunately, for this early Earth paradise, a 'Sword of Damocles' lay over its future and like a ticking time-bomb Earth's destiny was sealed from the very beginning.

There are many ancient historical accounts of the Great Catastrophe which came upon the Earth, its life forms and humankind; we know it today as the Great Deluge, the Biblical Flood.

The Epic of Gilgamesh

Gilgamesh was a king of Uruk, in Mesopotamia, who lived sometime between 2800 and 2500 BC. He is the main character in the Epic of Gilgamesh, an Akkadian poem that is considered the first great work of literature, and in earlier Sumerian poems where he is referred to as Bilgamesh.

The latest and most comprehensive telling of the Gilgamesh legend was the twelve-tablet Standard Babylonian Version, compiled circa 1200 BC.

In the epic, Gilgamesh is a demigod of superhuman strength who built the city walls of Uruk to defend his people and travelled to meet the sage Utnapishtim (Noah), who survived the Great Deluge. According to the Sumerian King List, Gilgamesh ruled his city for 126 years.

'In those days the world teemed, the people multiplied, the world bellowed like a wild bull, and the great god was aroused by the clamour. Enlil heard the clamour and he said to the gods in council, 'The uproar of mankind is intolerable and sleep is no longer possible by reason of the babel.' So the gods agreed to exterminate mankind. Enlil did this, but Ea because of his oath warned me in a dream. He whispered their words to my house of reeds, 'Reed-house, reed-house! Wall, O wall, hearken reed-house, wall reflect; O man of Shurrupak, son of Ubara-Tutu; tear down your house and build a boat, abandon possessions and look for life, despise worldly goods and save your soul alive. Tear down your house, I say, and build a boat. These are the measurements of the barque (the Ark) as you shall build her: let hex beam equal her length, let her deck be roofed like the vault that covers the abyss; then take up into the boat the seed of all living creatures.'

After Utnapishtim (Noah) builds the boat, the ensuing catastrophe is described:

(I have underlined important descriptive parts)

'The time was fulfilled, the evening came, the rider of the storm sent down the rain. I looked out at the weather and it was terrible, so I too boarded the boat and battened her down. All was now complete, the battening and the caulking; so I handed the tiller to Puzur-Amurri the steersman, with the navigation and the care of the whole boat. With the first light of dawn a black cloud came from the horizon; it thundered within where Adad, lord of the storm was riding. In front over hill and plain Shullat and Hanish, heralds of the storm, led on. Then the gods of the abyss rose up Nergal pulled out the dams of the nether waters (the waters below ground), Ninurta the war-lord threw down the dykes, and the seven judges of hell, the Annunaki, raised their torches, lighting the land with their livid flame. A stupor of despair went up to heaven when the god of the storm turned daylight to darkness, when he smashed the land like a cup.'

This historical event recorded by many cultures across the globe is described as a terrifying, total annihilation catastrophe falling upon humankind. The description continues:

'One whole day the tempest raged, gathering fury as it went, it poured over the people like the tides of battle; a man <u>could not see his brother</u>, nor the people be seen from heaven. Even the gods were <u>terrified at the flood</u>, they fled to the highest heaven, the firmament of Ann; they crouched against the walls, cowering like curs. Then Ishtar the sweet-voiced Queen of Heaven cried out like a woman in travail: 'Alas the <u>days of old are turned to dust</u> because I commanded evil; why did I command thus evil in the council of all the gods? I commanded <u>wars</u> to destroy the people, but are they not my people, for I brought them forth? Now like the spawn of fish <u>they float in the ocean</u>.' The great gods of heaven and of hell wept, they covered their mouths.'

This was no single local storm, but a global catastrophic event, the description is entwined with religious belief in order to try and interpret what was really happening to them and to the world:

'For <u>six days and six nights</u> the winds blew, torrent and tempest and <u>flood overwhelmed the world</u>, tempest and flood raged together like warring hosts. When the seventh day dawned <u>the storm from the south</u>[*] subsided, the sea grew calm, the, flood was stilled; I looked at the face of the world and there was silence, <u>all mankind</u> was turned to clay. The <u>surface of the sea stretched as flat as a roof-top</u>; I opened a hatch and the light fell on my face. Then I bowed low, I sat down and I wept, the tears streamed down my face, for <u>on every side</u> was the waste of water.' I <u>looked for land in vain</u>, but fourteen leagues distant there appeared a mountain, and there the boat grounded; on the mountain of Nisir[*] the boat held fast, she held fast and did not budge.'

Note^{**}: from the south would suggest the Arabian peninsula or the continent of Africa.

Note[**]: Mount Nisir is supposedly the mountain known as today as Pir Omar Gudrun (elevation 9000 ft.), near the city Sulaymaniyah in Iraqi Kurdistan. Mt. Nisir according to the Epic of Gilgamesh is the resting place of the ship built by Utnapishtim (Noah). Despite the precise descriptions in the Epic of Gilgamesh, the curious have never attempted to search for the remains of the giant ship (the Ark) on Mt. Nisir.

After the released birds finally find land Utnapishtim (Noah) makes an offering to the gods.

'Then, at last, Ishtar also came, she lifted <u>her necklace with the jewels of heaven</u> (a <u>rainbow</u>) that once Anu had made to please her. 'O you gods here present, by the lapis lazuli round my neck <u>I shall remember these days as I remember the jewels of my throat</u> (the <u>rainbow</u>); these last days I shall not forget. Let all the gods gather round the sacrifice, except Enlil. He shall not approach this offering, for without reflection he brought the flood; he consigned my people to destruction.'

The important physical descriptions of this global catastrophe may describe a combined geological and astronomical (catalytic) event.

The event is described as originating from the south (of Iraq), a line drawn south crosses the Arab Peninsula and into Africa. When I first read the Gilgamesh description it appeared to me as some kind of astronomical event, maybe a comet or asteroid impact on the African continent which sent explosive thunderous clashes through the heavens. An asteroid impact would certainly produce a similar description if observed from a distance and the explosive nature of such an impact would darken the skies with clouds of dust and debris and herald rain and thunderstorms. Such an asteroid impact would not bring such a great deluge of water from the heavens unless it was a giant icy comet, which may be the case, but in the text the descriptions of the nether waters of the abyss bursting from the ground is key to the amount of water needed for such a global deluge.

If we consider the untamed geological processes of early Earth forming vast reservoirs of water below the crust, it was inevitable that sooner or later the buildup of water pressure would reach a breaking point, thus bursting through, cracking the Earth's crust and releasing giant geysers of water which unleashed a global flood, as described by 'the land (Earth) was smashed like a cup.' An asteroid impact may have been the catalytic event which cracked an Earth under pressure. Once this enormous volume of primeval water was released into the atmosphere, Earth's temperature would drop; the water in the already saturated atmosphere would condense out as torrential rain and would be combined with the torrential flow of water from the abyss. With the Earth at this time being covered with shallow seas and low mountain ranges it would easily be swamped under such a deluge. The Earth's crust was no longer intact, but had been cracked into numerous pieces, what we see today as tectonic plates. The unleashing of nether water and torrential rain continued for several days until the internal pressures of the sub-crust water had dropped to almost zero.

The Earth had now entered a different geological time.

The enormous weight of water covering the Earth would push down on the already formed oceanic crusts, sending them deeper and deeper into the mantle, and at the same time giant mountain ranges would be pushed slowly upwards. The Genesis account records that the rains that created the Flood lasted for 40 days (Genesis 7:17), that the waters prevailed on the earth for 150 days (Genesis 7:24), and after these 150 days the waters gradually receded from the earth so that by the seventh month and the seventeenth day the Ark came to rest.

The post deluge appearance of the rainbow is also significant, for the Earth is no longer tropical, covered with the mists fed by out flowing geysers, clouds and rain droplets could now form, thus for the first time allowing a rainbow to form out of the clear sunlight.

20. Origins of Fossil Fuels

Fossil Rich Sedimentary Rocks and Mass Extinctions

After the discussion of the major catastrophic event described in 'The Day the Earth was Smashed like a Cup' it is little wonder that the geology of the Earth shows global sedimentary deposits rich in fossils. Such an astronomical event, a global deluge of unprecedented proportions from the bursting forth of the primeval mantle water, which had accumulated over geological time devastated all life on planet Earth:

The Biblical account in Genesis 7:11 records:

'...on that day all the <u>springs of the great deep</u> burst forth, and the <u>floodgates of the heavens</u> were opened. And rain fell on the earth forty days and forty nights.'

The order here recorded is as we described, first the water from the deep followed by the downpour of rain from a once water super saturated atmosphere.

Genesis 7:17-24 records:

'For forty days the <u>flood kept coming</u> on the earth, and as the waters increased they lifted the ark high above the earth. The waters rose and <u>increased greatly</u> on the earth, and the ark floated on the surface of the water. They rose greatly on the earth, and <u>all the high mountains</u> under the entire heavens were covered. The waters rose and <u>covered the mountains</u> to a depth of more than fifteen cubits. Every living thing that moved on land <u>perished</u> - birds, livestock, wild animals, <u>all</u> the creatures that swarm over the earth, and <u>all mankind</u>. Everything on dry land that had the breath of life in its nostrils died. Every living thing on the face of the earth was wiped out; people and animals and the creatures that move along the ground and the birds were wiped from the earth. Only Noah was left, and those with him in the ark. The waters flooded the earth for <u>a hundred and fifty days (5 months)</u>.'

There is little doubt that such a major catastrophe of the primeval mantle waters breaking forth from beneath Earth's crust would cause such devastation to life on Earth.

The waters continued to cover the Earth:

Genesis 8:5 records:

'The <u>waters continued to recede</u> until the <u>tenth month,</u> and on the first day of the tenth month the <u>tops of the mountains became visible.</u>

Genesis 8:13 - 14:

'By the first day of the first month of Noah's six hundred and first year, the <u>water had dried up from the earth</u>.... By the twenty-seventh day of the second month the earth was completely dry.'

It can be seen from the timeline that the waters covered the whole Earth for a whole year, during this time deep layers of mud and the remains of the living creatures would have settled out into sedimentary layers which covered the Earth.

There are accounts of mass extinctions of life on planet Earth, but none have been attributed to such a global catastrophe? The tropical age of giant creatures and pre-flood civilizations had been wiped out almost completely from the geological and historical record. It is recorded that God promised never to flood the Earth again and the rainbow symbolized that promise, but as we have seen from the geological formation of the Earth, these under pressure mantle waters once unleashed could never build up again, the majority of Earth's primordial water was now on the surface of the Earth and besides, Earth's crust having been 'smashed like a cup' was now open to gradual release of any remaining primeval mantle water.

The Origins of Oil, Coal and Gas

The dominant view of the origin of oil amongst western oil companies until 1969 was that it was due to the decay of living matter. Now other views are making themselves heard. To try and resolve the issue whether oil is biogenic (derived from living matter) or abiogenic (built up from primordial matter and therefore not from living matter) a Hedberg Conference recently took place. The issue was not resolved. A few of the interesting facts from the conference:

1. Hydrocarbons (a dominant part of petroleum) are clearly identified as break-down products of organic molecules from plants.

2. Hydrocarbons show preference for odd-numbered alkenes (which is considered to be a feature of 'organic oil'):

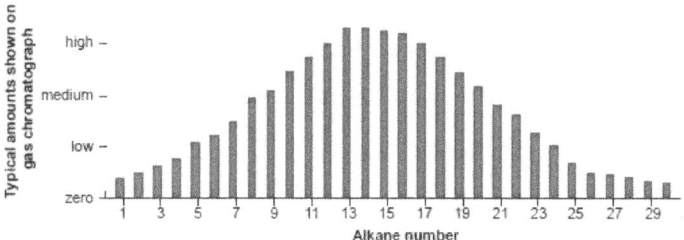

Hydrocarbons show optical activity. They rotate plane-polarized light indicating left-handed molecules are preponderant as found in living organisms. Hydrocarbons are found in sedimentary rocks rather than primary rocks.

Basic Oil Geology

Oil deposits are usually found in sedimentary rocks. Such rocks formed as sand, silt, and clay grains were eroded from land surfaces and carried by moving water to be deposited in sediment layers. As these sediment layers dried, chemicals from the water formed natural cements to bind the sediment grains into hard rocks.

Pools of oil are found in underground traps where the host sedimentary rock layers have been folded and/or faulted. The host sedimentary or reservoir rock is still porous enough for the oil to accumulate in spaces between the sediment grains. The oil usually hasn't formed in the reservoir rock but has been generated in source rock and subsequently migrated through the sedimentary rock layers until trapped.

Origins - Plant and Animal Decay

Most scientists agree that hydrocarbons (oil and natural gas) are of organic origin, where evidence shows that petroleum comes from plants (and animals), which were buried and fossilized in sedimentary source rocks. The petroleum was then chemically altered into crude oil and gas. The chemistry of oil provides crucial clues as to its origin. Petroleum is a complex mixture of organic compounds. One such chemical in crude oils is called porphyrin:

Porphyrins are organic molecules that are structurally very similar to both chlorophyll in plants and hemoglobin in animal blood. The porphyrin, Heme, is the pigment in red blood cells.

The presence of porphyrins in some petroleums means that anaerobic conditions developed early in the life of such petroleums, for chlorophyll derivatives, such as porphyrins, are easily and rapidly oxidized and decomposed under aerobic conditions. The fact that porphyrins are still present in crude oils today must mean that the petroleum source rocks and the plant (and animal) fossils in them had to have been kept from the presence of oxygen when they were deposited and buried: porphyrins break down under heat and in the presence of oxygen which suggests that:

1. The sedimentary rocks were deposited under oxygen deficient (or reducing) conditions.6

2. The sedimentary rocks were deposited so rapidly that no oxygen could destroy the porphyrins in the plant and animal fossils.7

Such environments are too rare to explain the presence of porphyrins in all the many petroleum deposits found around the world. The only consistent explanation is the catastrophic sedimentation that occurred during the worldwide Genesis Flood. Tons of vegetation and animals were violently uprooted and killed respectively, so that huge amounts of organic matter were buried so rapidly that the porphyrins in it were removed from the oxidizing agents which could have destroyed them. Crude oil porphyrin can be made from plant chlorophyll in less than 12 hours. Experiments have shown that plant porphyrin breaks down in as little as three days when exposed to temperatures of only 410°F (210°C) for only 12 hours. Therefore, the petroleum source rocks and the crude oils generated from them can't have been deeply buried to such temperatures for millions of years.

The Millions of Years Needed to Form Oil Misconception

Crude oils themselves <u>do not take long</u> to be generated from appropriate organic matter. Most petroleum geologists believe crude oils form mostly from plant material, such as diatoms (single-celled marine and freshwater photosynthetic organisms) and beds of coal (huge fossilized masses of plant debris). The latter is believed to be the source of most Australian crude oils and natural gas because coal beds are in the same sequences of sedimentary rock layers as the petroleum reservoir rocks.

Thus, for example, it has been demonstrated in the laboratory that moderate heating of the brown coals of the Gippsland Basin of Victoria, Australia, to simulate their rapid deeper burial, will generate crude oil and natural gas similar to that found in reservoir rocks offshore in only 2–5 days.* However, because porphyrins are also found in <u>animal blood,</u> it is possible some crude oils may have been <u>derived from the animals also buried and fossilized in many</u> <u>sedimentary rock layers</u>. Indeed, animal slaughterhouse wastes are now routinely converted within two hours into high-quality oil and high-calcium powdered and potent liquid fertilizers, in a commercial thermal conversion process plant

Conclusion

All the available evidence points to a recent catastrophic origin for the world's vast oil deposits, from plant and other organic debris, consistent with a global deluge. Vast forests grew on land and water surfaces in the pre-Flood world, and the oceans teemed with diatoms and other tiny photosynthetic organisms. Then during the global Flood cataclysm, the forests were uprooted and swept away. Huge masses of plant debris were rapidly buried in what thus became coal beds, and organic matter generally was dispersed throughout the many catastrophically deposited sedimentary rock layers. The coal beds and fossiliferous sediment layers became deeply buried as the Flood progressed. As a result, the temperatures in them increased sufficiently to rapidly generate crude oils and natural gas from the organic matter in them. These subsequently migrated until they were trapped in reservoir rocks and structures, thus accumulating to form today's oil and gas deposits.

It seems ironic that daily we burn in our cars, homes and industry the fossil evidence of a global catastrophic deluge.

* Animal Wastes Become Oil: Turkey and pig slaughterhouse wastes are daily trucked into the world's first biorefinery, a thermal conversion processing plant in Carthage, Missouri. On peak production days, 500 barrels of high-quality fuel oil better than crude oil are made from 270 tons of turkey guts and 20 tons of pig fat.

21. The Antediluvian (Pre-Deluge) Age of Old Kings and Patriarchs

One of the puzzling accounts of the pre-deluge Sumerian kings of Gilgamesh and the ancient Patriarchs of the Biblical Noah is their recorded longevity.

The Sumerian King List records the lengths of reigns of the kings of Sumer. The age of kings before the Flood is significantly different from that recorded after.

It is suggested that the Sumerian scribe that composed the original antediluvian list had available a document (possibly a clay tablet) containing numerical information on the ages of eight of the patriarchs similar to that of the Genesis record and that he mistakenly interpreted it as being written in the sexagesimal system (60^n):

$$\text{six } 10 \times 60^2 \text{ signs, six } 60^2 \text{ signs and six } 60^1 \text{ signs.}$$

The lives of the biblical patriarchs, however, have a precision of one year. If Adam and Noah are not included (as in the King List), and the lives of the patriarchs are similarly rounded to two digits, the sum of the lives has six 10^3 signs, six 10^2 signs and six 10^1 signs.

In addition, if the number representing the sum of the ages was wrongly assumed as having been written in the sexagesimal system, the two totals become numerically equivalent.

That the two documents are numerically related is strong evidence for the historicity of the book of Genesis. The Sumerian account shows up as a numerically rounded, incomplete version of the Genesis description.

The early chapters of the book of Genesis contain numerical information about the ages of the biblical patriarchs and their chronological relationships during the antediluvian world:

It can be seen from the timelines that Noah was born several years before the death of Adam, the first man. If these recorded antediluvian longevities are correct then during this time it would not make much sense to record historical events, for if Noah wanted to know anything about the early years of the life of the first men, he could simply go and talk to them. These longevities need explanation in terms of the physical events of the primeval waters Great Deluge. This we will do at the end of this topic, but first let us examine the records more closely.

The Sumerian King List contains an initial section that makes reference to the Flood and to Sumerian kings of extremely long reigns before the Flood. The antediluvian portion of the King List is very different from the biblical account. It only contains eight kings, while Genesis has ten patriarchs. The Sumerian list does not include the Sumerian First man of the Flood hero, so excluding Adam and Noah both contain the same number of patriarchs, eight.

The Sumerian list assigns an <u>average</u> reign duration of <u>30,150</u> years, with a total duration for the period of 241,200 years, compared to an average age of the biblical patriarchs of <u>858</u> years and a sum of 8575 years for their full lives.

If the number base of the Sumerian list is incorrect and is changed from sexagesimal to decimal, then the total of the durations of the kingdoms and the total of the ages of the patriarchs are numerically related and are equivalent.

Ages of the Pre-Deluge Biblical Patriarchs

It can be seen from the following graphs that the ages of the pre-deluge 10 generation patriarchs was about 900 years and that after the great Deluge the age of man dropped slowly over the next 12 generations to more familiar life spans, 100 years or less.

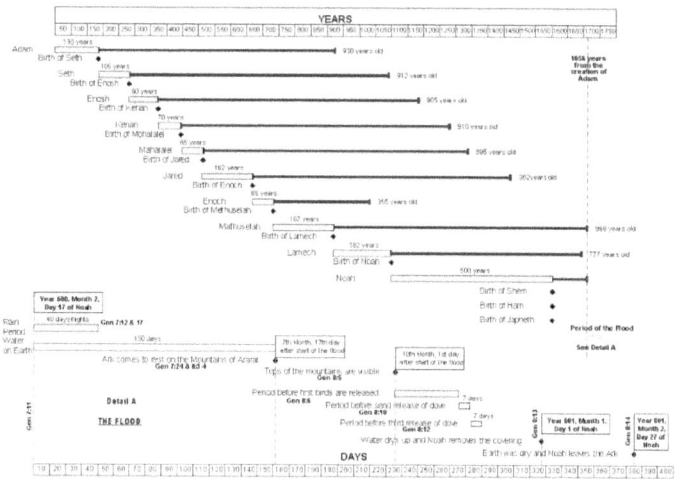

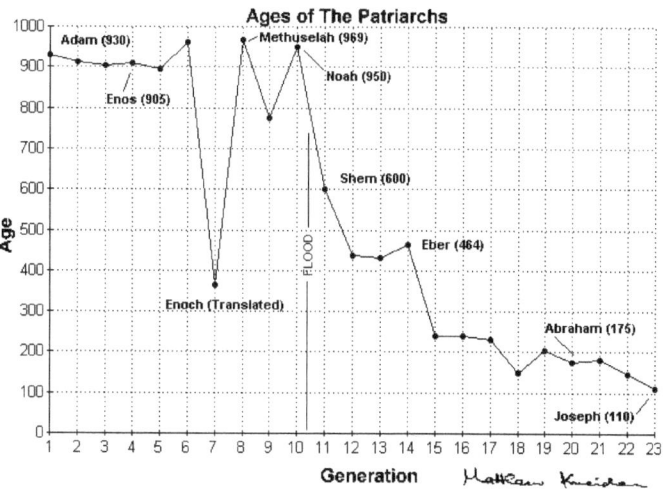

If we consider that before the Great Flood Earth was a warm tropical paradise of diverse plant life watered by the mists of the ground:

'...for the Lord God had <u>not caused it to rain</u> on the land, and there was no man to work the ground, and <u>a mist was going up from the land</u> and was <u>watering the whole face of the ground</u>.' Genesis2:5-6

This would have been a time of no direct sunlight or rain or rain clouds, and therefore no rainbows, if the sun during this early period gave a diffuse light, then the occupants of the Earth would have been protected from the harmful radiation (U.V. etc) of the sun. We know today that over exposure to the sun causes 'ageing' of the skin. UV-B, the shorter wavelength end of the U.V. spectrum causes damage at the molecular level to the fundamental building block of life, deoxyribonucleic acid (DNA) and can possibly cause genetic change. High doses of U.V. radiation is harmful to life.

Is it possible that post Deluge patriarchs because of exposure to direct sunlight were genetically changed to have shorter life spans, one thing for sure is that whatever the cause, history records that after the flood we lived shorter life spans.

Anti-ageing scientists today have identified a biological clock in the cells of living tissue and have in some experiments reversed this clock to produce young cells from old.

22. The Waters of Nun

In Ancient Egyptian creation accounts the original <u>mound of land comes forth from the waters of the Nun</u>. Nu 'watery one', also called Nun 'inert one' is the deification of the primordial watery abyss in ancient Egyptian religion.

niw, nww, nnw, nwnw. primeval waters, the mass of water which existed in primeval times, waters of chaos.

The ancient Egyptians were reminded of the original act of creation, the land emerging from the chaotic waters, annually, by the inundation of the Nile. Each year a fresh deposit of black fertile soil was brought upon the land and the emergence of new life in the form of shooting plants inspired the idea of the regenerative powers of the sacred mound, a mythological forerunner of burial mastabas and pyramids.

The following image portrays Nun carrying the sacred barque (boat, ark) with eight deities, the central deity represents Ra, the sun god.

This ancient Egyptian mythology of 8 deities being carried aloft out of the primeval waters of chaos may represent that of the Ark and the saved Deluvian eight, Noah and his wife, his three sons and their wives.

23. Göbekli Tepe - The Oldest Temple in the World

It has always been an unexplained enigma to me that for 200,000 years modern humans remained 'hunter gatherers' never settling down to agriculture or the construction of communal cities, towns and villages.

We have been led to believe that civilization originated only 5000 years ago with the rise of the ancient Egyptian and Sumerians along the Fertile Crescent of the rivers the Nile, Euphrates and Tigris. It was only a matter of time before this sterile, aseptic view of our origins was changed with new archaeological evidence, and this has come in the form of the new exciting discoveries of the ancient religious stone complex of Gobekli Tepe located at the top of a mountain ridge in the Southeastern Anatolia Region of Turkey, some <u>550 km south-west of Mount Ararat.</u>

The global deluge of primeval mantle water would have wiped out and or buried any signs of Antediluvian civilizations, what would remain on the residual surface of the land is primitive groups of peoples starting all over again, but with one difference they would have carried the knowledge the and skills of their pre-deluge ancestors.

Göbekli Tepe is a recently excavated site that was once buried under earth as would be expected for such an antediluvian site.

Göbekli Tepe (Turkish: 'Potbelly Hill') is an archaeological site at the top of a mountain ridge in the Southeastern Anatolia Region of Turkey, approximately 12 km northeast of the city of Şanlıurfa. The hill has a height of 15 m and is about 300m in diameter. It is approximately 760 m above sea level. It has been excavated by a German archaeological team that was under the direction of Klaus Schmidt from 1996 until his death in 2014.

The hill includes two phases of ritual use dating back to the <u>10th-8th millennium BC.</u>

During the first phase, circles of massive T-shaped stone pillars were erected. More than 200 pillars in about 20 circles are currently known through geophysical surveys. Each pillar has a height of up to 6 m and a weight of up to 20 tons. They are fitted into sockets that were hewn out of the bedrock. In the second phase, the erected pillars are smaller and stood in rectangular rooms with floors of polished lime. Topographic scans have revealed that other structures next to the hill, awaiting excavation, probably date to 14-15 thousand years ago, the dates of which potentially extend backwards in time.

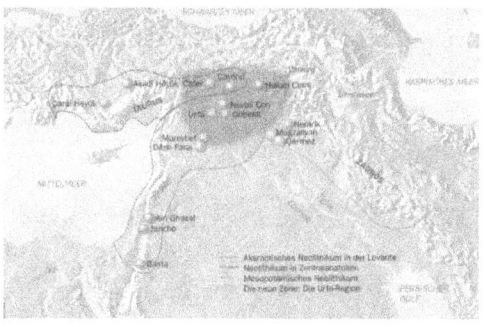

The site has been dated by excavated pottery and radiocarbon dating of charcoal in the lower layers to 7500-9100 BC, before the Great Deluge.

Göbekli Tepe is regarded as an archaeological discovery of the greatest importance since it could profoundly change the understanding of a crucial stage in the development of human society.

24. The Pre-Dynastic Glorious Period of Ancient Egypt
The Reign of Gods and Demigods

The ancient Egyptians recorded a Pre-Dynastic Period, what would be historically an Antediluvian period, the reign of their kings and the age of mythical god-kings.

The Sun reigned:	32,000 yrs
The Gods reigned:	984 yrs
The Demi-Gods reigned:	217 yrs
Total Glorious Period:	34,201 yrs
Dynastic Pharaoh Kings:	2324 yrs

Little is known of the Glorious Period of God-Kings of ancient Egypt's Pre-Dynastic remote past, but recorded legends of them survived into the Dynastic era. With the passing of time these early warring tribal kings became immortalized as gods. The King's List of Manetho and the Turin Cannon Kings List both commence with reigns of gods and Demi-gods:

Ra, Atu, Khepri, Horus, Seth, Osiris, Tefnut, Nut, Geb, Nephthys and Isis etc.

The legends of the conflict between the gods Horus, Osiris and Seth are probably based in the truth of actual tribal warfare. In Egyptian mythology the immediate descendants of these gods were the pre-dynastic kings:

Lower Egypt: Seka, Khaau, Tau, Thesh, Neheb, Wadjin and Menkhe

Upper Egypt: De, Ra, Ka, Sma, Scorpion, Narmer, Aha

The last three kings of Upper Egypt in the south became the early Dynastic kings who conquered the north and united Egypt and thus started the Pharaoh-King Dynastic Period. These three kings, Scorpion, Narmer and Aha may have been the First Pharaohs of a united Egypt and one or all of them may have been the legendary King Menes as recorded by Manetho* as the first pharaoh.

Note*: Manetho was an Egyptian priest from Heliopolis. Under the patronage of Ptolemy I, he started to compile an Egyptian history, titled Aegyptika. It was written in Greek and finished c.271 BCE. He attempted to describe Egyptian history from its inception to Alexander the Great. His original work was lost, however, and all we know is from short transcripts and summaries done by Christian historians in later centuries. A few different translations of his work exist:

Josephus Flavius, from the first century CE
Sextus Julius Africanus, third century CE
Eusebius of Cesarea, third/fourth century CE
George Syncellos (a Byzantine historian) from the eighth century CE.

25. King Menes - the First Post Deluge King of Egypt

As stated above King Menes would have been the first post Deluge king of Egypt and he is identified as being either: Scorpion, Narmer or Aha or all three may be one and the same. Traditionally, Narmer is associated with Manetho's Menes. A cosmetic palette bears his name:

The Narmer Palette, also known as the Great Hierakonpolis Palette or the Palette of Narmer, is a significant Egyptian archeological find, dating from about the 31st century BC, containing some of the earliest hieroglyphic inscriptions ever found.

It is thought by some to depict the unification of Upper and Lower Egypt under king Narmer. On one side, the king is depicted with the bulbous White Crown of Upper (Southern) Egypt, and the other side depicts the king wearing the level Red Crown of Lower (Northern) Egypt.

Along with the Scorpion Mace-head and the Narmer Maceheads, also found together in the Main Deposit at Nekhen, the Narmer Palette provides one of the earliest known depictions of an Egyptian king.

Narmer's hieroglyphic name is depicted in a Serekh (Royal Palace Facade) at the top middle of the palette and above and to the right of his head:

fish-chisel
nar-mr
fighting catfish - mr
'Narmer'

His name is composed of two hieroglyphs a *nar*-fish (fighting cat-fish) and a *mer*-chisel spelling: Nar-mer, however a fish hieroglyph is normally spelt: *an*, which would make the usual spelling of his name: *Anamer*.

Whichever is the correct phonetic spelling, Narmer, Anemer, Anamer or Anarmer, they are similar in consonantal sound.

The fish hieroglyph is found in the word:

Aha
fighting arms with shield and mace-letter A- fish ideogram
'Fighting Fish'

Here the fish sign it acts as an ideogram and the word is reminiscent of the name for King <u>Aha</u> which is depicted with the fighting sign in a Serekh and the Horus falcon:

Hor-Aha
'The Fighting Hawk'

This would suggest <u>Narmer</u> (<u>Anamer</u>) and King <u>Aha</u> are one and the same kings.

Since we assume Narmer was the first Post-Deluge Pharaoh in what way was he related to the survivors of the Deluge, since the world needed to be repopulated he must have been one of the descendants of Noah.

Descendants of Noah

In the Biblical account of the Great Deluge Catastrophe where the primeval mantle waters flooded the whole Earth, from his family came the descendant nations which re-populated the Earth.

Noah, Gilgamesh's Utnapishtim, had three sons, Shem, Ham and Japheth and from these four men and the wives came forth the seed of the nations.

Mizraim (Egypt)

Ham's son was <u>Mizraim</u> who by tradition settled in North Africa and populated the ancient Egyptian race.

Mizraim (Arabic, Miṣr) is the Hebrew and Aramaic name for the land of Egypt, with the dual suffix -āyim, perhaps referring to the 'two Egypts': Upper Egypt and Lower Egypt.

Neo-Babylonian texts use the term Mizraim for Egypt. The name was for instance inscribed in the famous Ishtar gate of Babylon. Ugaritic inscriptions refer to Egypt as Msrm, in the Amarna tablets it is called Misri, and Assyrian and Babylonian records called Egypt Musur and Musri.

Even today the Arabic word for Egypt is Misr (pronounced Masr in Egyptian colloquial Arabic).

According to Genesis 10, Mizraim was the younger brother of Cush and elder brother of Phut and Canaan, whose families together made up the Hamite branch of Noah's descendants.

Mizraim's Sons

Mizra-im (Msr: Egypt) had seven sons who with him settled in the Land of Egypt: Ludim, Anamim, Lehabim, Naphtuhim, Pathrusim, Casluhim, and Caphtorim.

According to Eusebius' Chronicon, Manetho had suggested that the great age of antiquity in which the later Egyptians boasted had actually preceded the flood, and that they were really descended from Mizraim, who settled there anew.

A similar story is related by medieval Islamic historians and the Persians, stating that the pyramids, etc. had been built by the wicked races before the deluge, but that Noah's descendant Mizraim was entrusted with reoccupying the region afterward.

According to George Syncellus, the Book of Sothis, supposedly by Manetho, had identified Mizra-im with the legendary first Pharaoh Menes, said to have unified the Old Kingdom and built Memphis. Mizraim also seems to correspond to Misor.

It seems therefore that the first early Dynastic king of Egypt was Mizraim (possibly known as the Scorpion king) and was the grandson of Noah.

One of the sons of Mizra-im is of particular importance for his name was <u>Anamim</u>, the sound of which is similar to the first true Pharaoh, <u>Narmer</u>, whose name can be translated as: An-mer, Nar-mer, Anamer.

If this is one and the same person, then the first true pharaoh of ancient Egypt was the great grandson of Noah, <u>Anamim</u>. It was he who united all the descendants of Mizraim. A text from Assyria, dating from the time of Sargon II, apparently calls the Egyptians 'Anami,' also known as the <u>Anamites</u>.

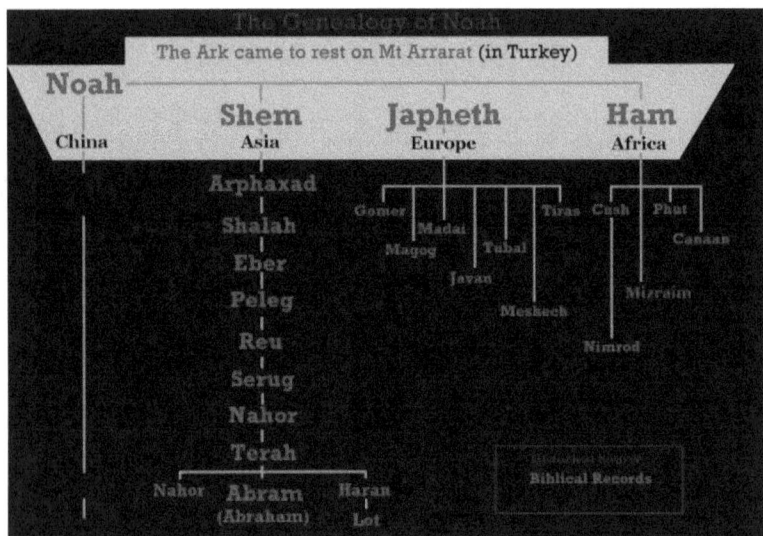

26. Thoth, the Giver of Writing God of the Divine Words - Hieroglyphs

The ancient hieroglyphic writing system appeared in Egypt with the first Dynastic Kings; arriving with the post Deluge descendants of Noah as an almost complete writing system from the start. According to the ancient Egyptians, the Medew-Netjer 'Divine Words' or 'Words of the God' was given to them by the god Thoth (Egyptian: Djhwty)

According to the Phoenician writer Sanchuniathon, Taautus (possibly Anamim) of Byblos, was the inventor of writing and son of Misor (Mizraim) who was bequeathed the land of Egypt by Cronus (Noah).

> '...Taautus (Djhwty) was the first who thought of the invention of letters, and began the writing of records...'

Djhwty was the god of Knowledge and Wisdom and since he gave the skill of writing in hieroglyphs to the first Egyptians it can be assumed that he was a man of literary skills and an early pharaoh who was later deified. If the Phoenician writer is correct that Djhwty was the son of Mizraim then we can assume that Thoth was the deified Narmer, Anamim.

27. Gilgamesh - The King of Sumer

When the flood waters receded to an acceptable level, Noah and his wife, Emzara disembarked from the Ark along with their sons, Shem, Ham, and Japheth, and their respective wives.

The descendants of Shem became the Shemites, or Semites (Semitic line of descent); the descendants of Japheth, the Indo-European nations, also known as the Gentiles; and the descendants of Ham, the Canaanites, Babylonians, Egyptians, and the Philistines.

Post Deluge, the grandson and great grandson of Noah, Mizraim (Egypt) and his son Anamim (Narmer), settled in North Africa and ruled and fathered the nation of Egypt.

Another great grandson of Noah, Nimrod, settled post deluge to the east. (Noah-Ham-Cush-Nimrod)

Nimrod king of Shinar, was, according to the Book of Genesis and Books of Chronicles, the son of Cush and great-grandson of Noah. He is depicted in the Bible as a mighty one in the earth and a mighty hunter. Extra-biblical traditions associating him with the Tower of Babel led to his reputation as a king who was rebellious against God. The association with the city of Erech (Babylonian Uruk) also attests the early provenance of the stories of Nimrod. Several Mesopotamian ruins were given Nimrod's name by 8th-century Arabs, including the ruins of the Assyrian city of Kalhu.

Nimrod rebelled against God for destroying the world with a Deluge:

Josephus wrote: 'Now it was Nimrod who excited them to such an affront and contempt of God. He was the grandson of Ham, the son of Noah, a bold man, and of great strength of hand. He persuaded them not to ascribe it to God, as if it were through his means they were happy, but to believe that it was their own courage which procured that happiness. He also gradually changed the government into tyranny, seeing no other way of turning men from the fear of God, but to bring them into a constant dependence on his power.

He also said he would be <u>revenged on God, if he should have a mind to drown the world again</u>; for that he would build a tower too high for the waters to reach. And that he would avenge himself on God for destroying their forefathers.'

As Nimrod's influence grew, he established the Cities of Erech, Nineveh, Babel, and Akkad among others, which would become the land of Shinar, or Sumer, the beginning of the kingdom of Babylonia. It has been suggested that Nimrod and Ninus (In Greek mythology, King of Assyria and founder of the city of Nineveh) was the same person. Even more interesting, theories have emerged, which indicate that <u>Nimrod</u> might have actually been <u>Gilgamesh</u>, the hero of a Babylonian epic, inscribed on ancient clay tablets, that parallels the Biblical story of Noah and the flood. According to the tablets, Gilgamesh was from Erech, a city attributed to Nimrod. Genesis 10:8-11, states that Nimrod established a kingdom. Since the Babylonian kingdom seems to be one of the earliest, if not the first kingdom on earth, it stands to reason that such an event would be recorded in extra-Biblical literature. And it was. Not only was the epic of Gilgamesh recorded on Sumerian tablets, but similar tales are found among the Assyrian and Hittite cultures as well.

Nimrod, who rebelled against God as being Gilgamesh makes sense, for we also read from his Epics that:

In the epic, Gilgamesh is a demigod of superhuman strength who built the city walls of Uruk to defend his people and travelled to meet the sage Utnapishtim (Noah), who survived the Great Deluge.

In other words, it seems that Nimrod (Gilgamesh, who worshipped pagan gods as part of his rebellion), visited his Great Grandfather, Noah (Utnapishtim).

28. Light and Water - The Origins of Life

In the beginning during the early stages of the formation of our universe, three simple atoms were fused together, two of hydrogen and one of oxygen, and formed one of the most remarkable molecules in our universe; the water molecule, which so many chemical and biochemical life processes depend upon.

Water is bound up in science and mythology, it is the great cleanser and out of water according to many creation myths and historical accounts did the Earth and Life come into existence.

Our early universe just after the Big Band was composed mostly of hydrogen, the simplest atom and of some helium, but before this when the universe was at its hottest and most dense stage there existed only pure light energy. This invisible to the naked eye light energy was in the form of extremely high frequency photons (wave-form packets), what science calls gamma rays. As the universe expanded and cooled, the light condensed into sub-atomic particles: protons, electrons and neutrinos (the three fundamental particles of matter). Further expansion and cooling allowed these particles to form simple atoms, mostly hydrogen which is composed of a single proton nucleus with an orbiting electron.

At this stage the universe was still dark, filled with dense clouds of an enormous amount of hydrogen. Under gravity, local cooler and therefore denser regions within the cloud contracted to form giant hot proto-stars. As the proto-star collapsed further under gravity, the temperatures and pressures in the core of the proto-star increased until it reached ignition point for nuclear fusion processes - a star was born. All over the entire universe billions of stars were being formed and igniting to give their light. The universe lit up in blinding flashes of light, like a giant firework display of enormous proportions.

Large mass stars, about 100 times the mass of our sun, burned brightly and quickly. These blue giant stars, because of their extreme core temperatures burned their hydrogen fuel at terrifying rates and in so doing fused together atomic nuclei to produce the heavier elements. At the end of their lifetime these stars imploded and exploded into supernova, and in this almost instantaneous big bang produced even heavier elements. The element rich clouds of these supernova stars provided the material for the formation of later generation stars and planetary systems. Gold is an element that requires the death of a star for its formation. These elements reacted together to form many of the chemical compounds found in the Earth.

Early Earth was abundant with the water compound of hydrogen and oxygen. Our vast oceans were essential for the ambient temperatures on our planet needed to support life. Water has a high heat capacity which means it heats up and cools down slowly ensuring that our Earth has even climatic temperatures without the extremes.

Another unusual property of water which is that as it freezes and turns to ice, it expands, making the ice less dense that liquid water, unlike most substances which when turning from liquid to solid become more dense. This strange property of water allows ice to float on liquid water, if the reverse were true our seas and lakes would freeze from the bottom upwards making them inhabitable for life.

In biological chemical processes water molecules are involved at almost every stage, making it the chemical of life. We can live for weeks without food, but only days without water.

29. Abiogenesis - The Generation of Life from Simple Organic Chemicals

If we imagine early sterile Earth of seas and soils rich in minerals, an ambient climate and an abundance of water, then it seems we have all the ingredients for life to spontaneously emerge in the form of simple replicating units. This emergence of life from simple inorganic chemicals is the very foundation stone of evolution, if life can evolve, then it will.

The big question, therefore is, can life evolve from the basic ingredients from which it is made?

Unfortunately, life forms, even the simplest, are extremely complex. Simple living single cell life forms are made of complex organic compounds which are involved in complex biochemical reactions and processes - biochemistry. There is no such thing as a 'simple' life-form. A scientist who studies such life forms and their biochemical and genetic makeup recently stated that the gulf between simple organic chemicals and simple life forms is enormous and getting wider the more we realize the complexity of their biochemistry. Nevertheless, the study of Abiogenesis is an important science to the foundation of evolution:

Abiogenesis

Abiogenesis is the natural process of life arising from non-living matter, such as simple organic compounds. The study of abiogenesis involves three main types of considerations: the geophysical, the chemical, and the biological.

Many approaches investigate how self-replicating molecules, or their components, came into existence. It is generally accepted that current life on Earth descended from an RNA world, although RNA-based life may not have been the first life to have existed. RNA is a DNA messenger molecule for making cell proteins. The Miller–Urey experiment and similar experiments demonstrated that most amino acids, basic chemicals of life, can be synthesized from inorganic compounds in conditions intended to be similar to early Earth.

Several mechanisms have been investigated, including lightning and radiation. Other approaches focus on understanding how catalysis in chemical systems in the early Earth might have provided the precursor molecules necessary for self-replication. 'Complex' organic molecules have been found in the solar system and in interstellar space, and these molecules may have provided starting material for the development of life on Earth. These molecules described as 'complex' are much simpler than the real complex molecules of life such as the protein enzymes which govern and mediate the chemical reactions in cells.

According to the panspermia hypothesis, microscopic life is distributed by meteoroids, asteroids and other small Solar System bodies, but this just shifts the problem somewhere else, if life can be generated under the most favorable conditions of early Earth, then it can happen on other similar planets.

Earth is the only place in the universe known to harbor life. The age of the Earth is about 4.5 billion years. The earliest undisputed evidence of life on Earth dates at least from 3.5 billion years ago according to modern dating methods.

The Complexity of Cellular Life

Most living things are composed of different kinds of cells specialized to perform different functions. A liver cell, for example, does not have the same biochemical duties as a nerve cell. Yet every cell of an organism has the same set of genetic instructions, so how can different types of cells have such different structures and biochemical functions? Since biochemical function is determined largely by specific enzymes (proteins), different sets of genes must be turned on and off in the various cell types. This is how cells differentiate.

This notion of cell-specific expression of genes is upheld by hybridization experiments that can identify the unique mRNAs in a cell type. More recently, DNA arrays and gene chips offer the opportunity to rapidly screen all gene activity of an organism. Co-expression of genes in response to external factors can thus be explored and tested.

DNA is not the same in every cell of the body: A new research reveals the complexity of cellular biochemistry. The research calls into question one of the most basic assumptions of human genetics: that when it comes to DNA, every cell in the body is essentially identical to every other cell. This discovery may undercut the rationale behind numerous large-scale genetic studies conducted over the last 15 years.

Until we can synthesize simple single cellular life-forms in sophisticated laboratories with modern science, it seems a very tall measure to expect a primordial pool of mud to form life.

30. Radioactive Dating and the Recently Discovered Variability of Decay Constants

Many accept radiometric dating methods as proof that the earth is millions of years old, in contrast to the biblical chronology of thousands of years.

The presupposition of long ages is an icon and foundational to the evolutionary model. Nearly every textbook and media journal teaches that the earth is billions of years old.

Using radioactive dating, scientists have determined that the Earth is about 4.5 billion years old, ancient enough for all species to have been formed through evolution.1 The earth is now regarded as between 4.5 and 4.6 billion years old.

The primary dating method scientists use for determining the age of the earth is radioisotope dating. Proponents of evolution publicize radioisotope dating as a reliable and consistent method for obtaining absolute ages of rocks and the age of the earth.

Radioactive dating is a technique used to date materials such as rocks or carbon, usually based on a comparison between the observed abundance of a naturally occurring radioactive isotope and its decay products, using known decay rates. The values of the decay constants used for radioactive dating for much of the last century have been set in stone as being true constants, unchanged by temperature of pressure, therefore a solid foundation for all dating of archaeological sites and rocks, including the age of the Earth itself. Recent scientific evidence has thrown it all into doubt and some confusion, namely, that these constants may not be constant at all.

There is also another stumbling block with radioactive dating; it assumes that we know how much radioactivity was originally present in the material when formed.

There are several assumptions that must be made in radioisotope dating. Three critical assumptions can affect the results during radioisotope dating:

1. The initial conditions of the rock sample are accurately known.

2. The amount of parent or daughter elements in a sample has not been altered by processes other than radioactive decay.

3. The decay rate (or half-life) of the parent isotope has remained constant since the rock was formed.

Radiocarbon Dating

This is an important method for dating organic carbon type materials, such as wood and cloth, found at archaeological sites.

Radiocarbon dating (carbon-14 dating) is a method of determining the age of an object containing organic material by using the properties of radiocarbon-14, a radioactive isotope of carbon. The method was invented in the late 1940s and soon became a standard tool for archaeologists. The radiocarbon dating method is based on the fact that radiocarbon is constantly being created in the atmosphere by the interaction of cosmic rays with atmospheric nitrogen. The resulting radiocarbon combines with atmospheric oxygen to form radioactive carbon dioxide (CO_2), which is incorporated into plants by photosynthesis; animals then acquire carbon-14 by eating the plants. When the animal or plant dies, it stops exchanging carbon with its environment, and from that point onwards the amount of carbon-14 it contains begins to reduce as the carbon-14 undergoes radioactive decay. Measuring the amount of carbon-14 in a sample from a dead plant or animal such as piece of wood or a fragment of bone provides information that can be used to calculate when the animal or plant died. The older a sample is, the less carbon-14 there is to be detected, and because the half-life (decay constant - the period of time after which half of a given sample will have decayed) is about 5,730 years, the oldest dates that can be reliably measured by radiocarbon dating are around 50,000 years ago, although special preparation methods occasionally permit dating of older samples.

This method <u>assumes</u> that the amount of carbon-14 originally present in the atmosphere during the time of the growth of the plant is known, it is based on an assumption of today's levels.

Because of the variation of Carbon-14 in the atmosphere in later years this method has been refined with the use of data from tree-rings extending the method back to 13,900 years.

Radioactive Dating of Rocks and the Earth

Radioactive elements were incorporated into the Earth when the Solar System formed. All rocks and minerals contain tiny amounts of these radioactive elements. Radioactive elements are unstable; they breakdown spontaneously into more stable atoms over time, a process known as radioactive decay. Radioactive decay occurs at a constant rate, specific to each radioactive isotope. Since the 1950s, geologists have used radioactive elements as natural 'clocks' for determining numerical ages of certain types of rocks.

Radiometric clocks are 'set' when each rock forms. 'Forms' means the moment an igneous rock solidifies from magma, a sedimentary rock layer is deposited, or a rock heated by metamorphism cools off. It's this resetting process that gives us the ability to date rocks that formed at different times in earth history.

A commonly used radiometric dating technique relies on the breakdown of potassium ($40K$) to argon ($40Ar$). In igneous rocks, the potassium-argon 'clock' is set the moment the rock first crystallizes from magma. Precise measurements of the amount of $40K$ relative to $40Ar$ in an igneous rock can tell us the amount of time that has passed since the rock crystallized. If an igneous or other rock is metamorphosed, its radiometric clock is reset, and potassium-argon measurements can be used to tell the number of years that has passed since metamorphism.

Radiometric dating has been used to determine the ages of the Earth, Moon, meteorites, ages of fossils, including early man, timing of glaciations, ages of mineral deposits, recurrence rates of earthquakes and volcanic eruptions, the history of reversals of Earth's magnetic field, and many of other geological events and processes.

The Variability of Radioactive Decay Constants

The story begins in classrooms around the world, where students are taught that the rate of decay of a specific radioactive material is a constant, no matter what temperature or pressure a radioactive substance is subjected to. This concept is relied upon, for example, when anthropologists use carbon-14 to date ancient artifacts.

Recent research by physicists has suggested that there is some correlation between changes in solar activity and radioactive decay rates. Scientists have found that there appears to be a correlation between the radioactive decay rate of Silicon-32 and Radium-226 on the earth and changes relating to the sun's activity. This is an important area of research for those who question the constancy of radioactive decay rates, and such variable decay rates may have a bearing upon our understanding of the dating of various rock layers.

Discrepancies in the radioactive decay of Silicon-32 and Radiu-226 on the earth's surface seem to show a degree of correlation with the annual cycle of the earth's orbit around the sun; with rates speeding up as the earth gets closer to the sun and decreasing as it moves away. This seems to correlate with a small time lag, or phase shift, between the distance between the sun and earth, and is as yet unexplained. Of course the changes observed in decay rates is small, perhaps of the order of less than 1 percent over the annual cycle, but then any changes in the distance from Earth to the sun is also relatively small.

In another study a sample of Manganese-54 was found to vary with the occurrence of a significant X-ray solar flare and a high-energy solar proton storm. It is noteworthy that the observed decay rate seems to have reduced with the flare, whereas the decay rate in the previous findings increased with closer proximity to the sun.

Research at Stanford University has also found that there is evidence of a correlation with the 33 day period spin of the solar core and decay rates of Silicon-32 and Chlorine-36. Research pointed out that one side of the sun's inner core emits neutrinos more strongly than the other, and from this accumulating evidence, it has been proposed that changes in neutrino flux differences may have some impact upon the rate at which some radioactive isotopes decay. Neutrinos as the cause of variability in decay constants has not been confirmed and the effects may be due to some unknown particle emitted from the sun.

An alternative explanation might be related to differences in the energy field density in the vacuum of space (sometimes called zero point energy observed as the Casimir effect, which is also related to theories of quantum gravity).

This recent work provides a glimpse into an area of research that may provide fruitful outcomes for those with an interest in creation science, particularly the variability of radioactive decay rates. Although at present the identified size of changes are quite small, there are perhaps early observational indications that larger changes in decay rates are possible.

Evidence that Radioactive Dating Can Be Wrong

We know that radioisotope dating does not always work because we can test it on rocks of known age. In 1997, a team of eight research scientists known as the RATE group (Radioisotopes and the Age of The Earth) set out to investigate the assumptions commonly made in standard radioisotope dating practices. Their findings were significant and directly impact the evolutionary dates of millions of years.

A rock sample from the newly formed 1986 lava dome from Mount St. Helens was dated using Potassium-Argon dating. The newly formed rock gave ages for the different minerals in it of between 0.5 and 2.8 million years. These dates show that significant argon (daughter element) was present when the rock solidified (assumption 1: false).

Mount Ngauruhoe is located on the North Island of New Zealand and is one of the country's most active volcanoes. Eleven samples were taken from solidified lava and dated. These rocks are known to have formed from eruptions in 1949, 1954, and 1975. The rock samples were sent to a respected commercial laboratory. The 'ages' of the rocks ranged from 0.27 to 3.5 million years old. Because these rocks are known to be less than 70 years old, it is apparent that assumption (1) is again false. When radioisotope dating fails to give accurate dates on rocks of known age, why should we trust it for rocks of unknown age? In each case the ages of the rocks were very greatly inflated.

Different methods of radioactive dating yield different ages and there are variations with the same method, so how can scientists know for sure the age of any rock or the age of the earth. In some cases the range of age for dated rocks was more than 500 million years.

Accelerated Decay

Recent studies have shown that sometime in Earth's past radioactive decay rates accelerated. Experiments the RATE project commissioned have clearly confirmed the numerical predictions of a young Earth model. The data and analysis showed that over a billion years worth of nuclear decay occurred very recently, between 4000 and 8000 years ago.

Other studies by the RATE group have provided evidence that radioactive decay supports a young earth. One of their studies involved the amount of helium found in granite rocks. Granite contains tiny zircon crystals, which contain radioactive uranium ($238U$), which decays into lead ($206Pb$). During this process, for each atom of $238U$ decaying into $206Pb$, eight helium atoms are formed and migrate out of the zircons and granite rapidly.

Within the zircon crystals, any helium atoms generated by nuclear decay in the distant past should have long ago migrated outward and escaped from these crystals. One would expect the helium gas to eventually diffuse upward out of the ground and then disappear into the atmosphere. To everyone's surprise, however, large amounts of helium have been found trapped inside zircons.

Based on the measured helium retention, a statistical analysis gives an estimated age for the zircons of 6,000 ± 2,000 years.

It must also be concluded, therefore, that because nuclear decay has been shown to have occurred at grossly accelerated rates when molten rocks were forming, crystallizing and cooling, the radiometric methods cannot possibly date these rocks accurately based on the false assumption of constant decay through earth history at today's slow rates. Thus the radiometric dating methods are highly unreliable and don't prove the earth is old.

Final Word

Whatever our origins, today we see an amazing complex and wondrous universe, full of the incredible diversity of life, from the smallest living single celled protozoa to the mighty majesty of an African elephant. The laws of the universe are so finely tuned and entwined as to make everything possible. We say our universe has evolved into such a remarkable thing, its harmony and complexity is beyond description, but that which is more incredible, is that from the beginning what we think were simple atoms, time, space energy and light, had the potential, the inbuilt laws and structure to produce everything we see to today, even sentient life produced by the universe can look back at itself and wonder. The universe is no more complex today, than when it first started, it is just unfolding.

http://arkpublishing.co.uk/

www.ingramcontent.com/pod-product-compliance
Lightning Source LLC
Chambersburg PA
CBHW070819180526
45168CB00002B/682